BookWare Companion Series ™

Discrete Systems Laboratory
Using MATLAB®

Books in the BookWare Companion Series^{TM}

BookWare Companion Series ™

Discrete Systems Laboratory
Using MATLAB®

Martin Schetzen

Vinay K. Ingle
Northeastern University

Australia • Canada • Mexico • Singapore • Spain • United Kingdom • United States

Publisher: Bill Stenquist
Sponsoring Editor: Heather Woods
Marketing Manager: Nathan Wilbur
Marketing Assistant: Christina DeVeto
Editorial Assistants: Shelley Gesicki, Meg Weist
Production Editor: Laurel Jackson
Cover Design: Denise Davidson
Print Buyer: Vena Dyer
Printing and Binding: Webcom Ltd.

MATLAB and Simulink are registered trademarks of The MathWorks, Inc. Further information about MATLAB, Simulink, and related publications may be obtained from The MathWorks, Inc., 3 Apple Hill Drive, Natick, MA 01760. E-mail: www.info@mathworks.com, http://www.mathworks.com; Phone: (508) 647-7001.

For more information, contact:
BROOKS/COLE
511 Forest Lodge Road
Pacific Grove, CA 93950 USA
www.brookscole.com

For permission to use material from this work, contact us by
Web: www.thomsonrights.com
fax: 1-800-730-2215
phone: 1-800-730-2214

Printed in Canada

10   9   8   7   6   5   4   3   2   1

Library of Congress Cataloging-in-Publication Data
Schetzen, Martin.
    Discrete Systems Laboratory Using MATLAB / Martin Schetzen, Vinay K. Ingle.
        p.   cm. – (BookWare companion series)
    Includes bibliographical references
    ISBN 0-534-37463-8 (alk. paper)
    1. Analog-to-digital converters – Mathematical models. 2. Digital-to-analog converters – Mathematical models. 3 Discrete-time systems. 4. Signal processing – Digital techniques. I. Ingle, Vinay K. II. Title II.

TK7887.6 .S346 2000
621.3815'32—dc21   99-047989

About the Series

> **"The purpose of computing is insight, not numbers."**
> —R.W. Hamming, *Numerical Methods for Engineers and Scientists,* McGraw-Hill, Inc.

It is with this spirit in mind that we present the BookWare Companion Series.™

Increasingly, the latest technologies and modern methods are crammed into courses already dense with important theory. The question is asked: "Are we simply teaching students the latest technology, or are we teaching them to reason?" We believe these two alternatives need not be mutually exclusive. This series was founded on the belief that computer solutions and theory can be mutually reinforcing. Properly applied, computing can illuminate theory and help students to think, analyze, and reason in meaningful ways. It can also help them understand the relationships and connections between new information and existing knowledge; and cultivate in them problem-solving skills, intuition, and critical thinking. The BookWare Companion Series was developed in response to this mission.

Specifically, the series is designed for educators who want to integrate their curriculum with computer-based learning tools, as well as for students who wish to go further than their textbook alone allows. The former will find in the series the means by which to use powerful software tools to support their course activities, without having to customize the applications themselves. The latter will find relevant problems and examples quickly and easily and will have electronic access to them. Important for both educators and students is the premise upon which the series is based—that students learn best when they are actively involved in their own learning. This series will engage them, provide a taste of real-life issues, demonstrate clear techniques for solving real problems, and challenge students to understand and apply these techniques on their own. The books in the series all encourage active learning.

To serve your needs better, we plan to improve the series continually. Join us at our BookWare Companion Resource Center web site (http://www.brookscole.com/engineering/ee/bookware.htm) and let us know how to improve the series; share your ideas on using technology in the classroom with your colleagues; suggest a great problem or example for the next edition; let us know what is on your mind. We look forward to hearing from you, and we thank you for your continuing support.

Heather Woods	Acquisitions Editor	heather.woods@brookscole.com
Shelley Gesicki	Editorial Assistant	shelley.gesicki@brookscole.com
Nathan Wilbur	Marketing Manager	nathan.wilbur@brookscole.com
Christina De Veto	Marketing Assistant	christina.deveto@brookscole.com

CONTENTS

Experiment 16
EFFECT OF FILTER REALIZATION ON THE OUTPUT MULTIPLICATION QUANTIZATION ERROR   112

PREFACE

Many physical processes are inherently discrete processes that can be modeled as discrete systems. Some common examples are population dynamics, business merchandise inventory, economic systems, and computer communication systems. In addition, many important systems are built as discrete systems. Examples include high-definition digital television, cellular phone systems, internet components, and digital control systems. The advent of generally available advanced digital computer technology has also stimulated the development of theoretical and applied techniques to model and study continuous processes with the aid of the digital computer. A block diagram of the process by which this is accomplished is shown below. As shown, the continuous waveform $s(t)$ first is sampled and quantized to convert it into a sequence of discrete values, $x(n)$, which can be processed by a digital computer. The system that accomplishes this conversion is called an analog-to-digital (or simply A/D) converter. The specific desired processing of the discrete sequence $x(n)$ to form another sequence, $y(n)$, is accomplished by a discrete system; this often is implemented by computer software or hardware. The sequence $y(n)$ then is converted to a continuous waveform, $r(t)$, by a digital-to-analog (or simply D/A) converter. Because D/A and A/D converters are inherently part of such systems, their study usually is included in courses on discrete systems and digital signal processing.

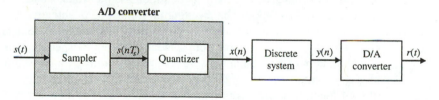

The experiments in this text are divided into four chapters. The experiments in Chapter I concern various aspects of sampling and D/A conversion. A/D quantization is not included in this chapter; rather, it is covered in Chapter IV, which is specifically concerned with various types of quantization errors. The focus of Chapter II is the Discrete Time Fourier Transform (DTFT), which is important in the study of linear, shift-invariant (LSI) discrete systems in the frequency domain. Then, in Chapter III, we use the frequency domain viewpoint to analyze and design basic types of digital filters.

Each experiment in this text is followed by a section of screen illustrations, which contains a description of each screen that appears on the monitor and some helpful guides for running the program. The screen-illustration section of an experiment should be read at least during the first time the experiment is run. The software is written for any computing platform running MATLAB version 5.0 or later.

The experiments in this text are designed to help students develop a logically accurate physical understanding of the basic concepts in digital signal processing and discrete systems. Some of the experiments are too long for a student to do properly in one lab session; however, we have divided those experiments into parts from which the lab instructor can choose an appropriate subset. The lab thus can be changed each term by choosing different experiments from each chapter and different parts of the long experiments.

It has been our experience in teaching discrete system theory that students often do not grasp the full physical significance of many mathematical results discussed in class. In consequence, the discrete systems laboratory we developed at Northeastern University is designed for students to take concurrently with the lecture course. Students are not required to do any programming in this laboratory. Our objective in developing these experiments was not only to demonstrate a theoretical result, but also to give students an opportunity to do some "original" studies by analyzing a given physical situation that has been simulated on the computer. Thus, several parts of the lab experiments are designed for students to consider themselves experimental investigators of a particular phenomenon. This often requires students to analyze some basic scientific concepts that often have been accepted without much concern. For example, the importance and use of dimensionless quantities is illustrated in Experiment 6. In Part 1 of Experiment 7, students learn that there are many possible definitions of *optimum*, so the optimum choice depends on the definition used. Further, students learn that for a criterion in science to be meaningful, it must be quantitative and that the definition of size must have certain properties. The interested students in our classes found these departures from standard cookbook experiments intellectually stimulating. These students were rewarded not only by a better grasp of the physical significance of the derived results and a good understanding of some of their theoretical and physical implications, but also by a greater appreciation of the scientific method and the interplay between theoretical and experimental studies.

This Bookware Companion book, *Discrete Systems Laboratory*, is intended to be used for a laboratory component of an undergraduate course on discrete systems and/or on digital signal processing. Since it has been written to accompany any course text, we have not included much of the theoretical development of the concepts used. However, we developed some concepts that class texts often do not include or do not present in a manner useful for the experimental study. We used the first five chapters of the text *Digital Signal Processing* by J. G. Proakis and D. G. Manolakis, as the class text. The students found this text to be an excellent complement for the class lectures and the laboratory experiments.

ACKNOWLEDGMENTS

We thank Prof. John G. Proakis, who was chairman of the Electrical and Computer Engineering Department of Northeastern University during the development and writing of these laboratory experiments. He provided the atmosphere and encouragement that made this project possible. We also thank Ms. Naomi Bulock of MathWorks, Inc., who was always helpful in providing the newest versions of MATLAB throughout this project.

   In no less degree, we are grateful to our wives, Jeannine Schetzen and Usha Ingle, whose patience and support have been indispensable.

Martin Schetzen
Vinay K. Ingle

CHAPTER I

SAMPLING AND RECONSTRUCTION

The experiments in Chapter I are concerned with a study of sampling. As we discussed in the text preface, an A/D converter changes a continuous waveform, $s(t)$, into a sequence of discrete values, $x(n)$, which can be processed by a digital computer. A/D conversion can be modeled as shown below.

A/D Converter

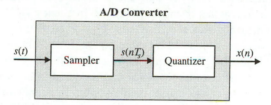

As shown in the diagram, the A/D converter is modeled as the tandem connection of two systems: a sampler and a quantizer. For the input $s(t)$, the sampler output is $s(nT_s)$, in which n is an integer and T_s is the sampling interval; that is, the sampler output is the sequence of values of the input at the time instances nT_s. Since a computer can recognize only a maximum number, L, of different numerical values determined by the computer word length, each value of the sampler output sequence must be quantized into one value from a set of L different values. This task is accomplished by the quantizer for which the output is $x(n)$ for the input $s(nT_s)$. The tandem connection of the sampler and the quantizer is a model of an A/D converter. The sampler is a linear, time-varying (LTV) system whose basic properties are studied experimentally in Chapter I. The quantizer is a time-invariant, nonlinear, no-memory system whose basic properties are studied experimentally in Chapter IV.

The basic phenomenon associated with sampling that will be experimentally examined in Chapter I is aliasing. Consider a sequence formed from samples of the sinusoid

$$s(t) = A \cos[2\pi F t + \phi] \tag{I.1}$$

taken at the time instances $t_n = nT_s$. There is no loss of generality incurred by always considering the sampling instances to be at the time $t_n = nT_s$ because the waveform being sampled can be shifted. Note that the sampling rate is $F_s = 1/T_s$ samples/second. The resulting sequence is

$$s(nT_s) = A \cos[2\pi F n T_s + \phi]$$

$$= A \cos[2\pi f n + \phi] \tag{I.2}$$

in which f, called the relative frequency, is $f = FT_s = F/F_s$ cycles per sample. Now, if $f = f_0 + k$ in which k is an integer, then

$$s(nT_s) = A \cos[2\pi(f_0 + k)n + \phi]$$

$$= A \cos[2\pi f_0 n + \phi] \tag{I.3}$$

The last equation follows since k and n are integers and $\cos(\theta + 2kn\pi) = \cos(\theta)$. From eqs. (I.2) and (I.3), we note that the same sequence would be obtained if the relative frequency were $f_0 = F_0/F_s$. That is, sequences whose relative frequencies differ by an integer are identical. Equivalently, the same sequence is obtained whether a sinusoid with the frequency $F_0 = F_s f_0$ hertz or a sinusoid with the frequency $F = F_s f = F_s(f_0 + k) = F_0 + kF_s$ hertz is sampled. That is, the sequences obtained by sampling sinusoids whose frequencies differ by kF_s are identical. This is called aliasing. In D/A conversion, the relative frequency, f, of any sinusoidal sequence is considered to be in the unit interval $-0.5 < f < 0.5$. In consequence, the maximum frequency of a sinusoid reconstructed from a sinusoidal sequence will never exceed $F_s/2$. Thus, if aliasing is to be avoided, we require $F_s > 2F$ so that the sampling rate, F_s, exceeds twice the frequency, F, of the sinusoid being sampled. Now, consider the waveform to be sampled, $s(t)$, to be the linear combination of sinusoids as

$$s(t) = \sum_k A_k \cos(2\pi F_s t + \phi_k) \tag{I.4}$$

In accordance with our discussion, the sampled sequence obtained then is

$$s(nT_s) = \sum_k A_k \cos(2\pi f_k n + \phi_k) \tag{I.5}$$

in which $f_k = F_k/F_s$. Also, observe that it is necessary that $|f_k| < \frac{1}{2}$ for each value of k so that there is no aliasing. Thus, it is necessary that $F_s > 2F_k$ for each value of k. Equivalently, it is necessary that $F_s > 2F_{max}$, in which F_{max} is the highest frequency of any sinusoid of which $s(t)$ is composed. Often, $2F_{max}$ is called the Nyquist rate, F_n. The condition for no aliasing then is that the sampling rate, F_s, must be larger than the Nyquist rate, F_n. Various aspects and consequences of aliasing will be explored experimentally in this chapter.

Experiment 1

IDEAL SAMPLING AND RECONSTRUCTION

BACKGROUND

The main objective of this experiment is to examine some of the basic concepts of sampling theory. In accordance with the sampling theorem, a bandlimited waveform, $s(t)$, with the highest frequency F_{max} hertz that is sampled at the rate F_s samples/second can be reconstructed without error from its sample values $s(nT_s)$ if $F_s = 1/T_s > 2F_{max}$.

A reconstructed waveform, $r(t)$, can be obtained by ideal interpolation using the sample values as follows:[1]

$$r(t) = \sum_{n=-\infty}^{\infty} s(nT_s)g(t - nT_s) \tag{1.1}$$

in which the interpolation function is

$$g(t) = \frac{\sin(\pi F_s t)}{\pi F_s t} \tag{1.2}$$

The sampling theorem states that $r(t) = s(t)$ if $F_s > 2F_{max}$. The reconstructed waveform, $r(t)$, is not necessarily equal to $s(t)$ if $F_s \leq 2F_{max}$ for then aliasing occurs. A D/A converter with the input sample values, $s(nT_s)$, and the output waveform, $r(t)$, obtained in accordance with eq. (1.1) is called an ideal D/A converter.

---

[1] The sampling theorem as stated above was first published in 1915 by the British mathematician E. T. Wittaker in the article, "On the Functions which are Represented by the Expansions of the Interpolation Theory," *Proc. Royal Society*, Edinburgh, Vol. 35, pp. 181–194. The engineer H. Nyquist was among the first Americans to apply this sampling theorem to practical problems. In recognition of Nyquist's applications and interpretations of this theorem, his name often is associated with it.

■ QUESTION 1.1   Show that a sampler is a no-memory, linear, time-varying system.

■ QUESTION 1.2   Show that an ideal D/A converter is a noncausal, linear, time-invariant system.

Note that the sampler and the ideal D/A converter satisfy superposition because they are linear systems. The software provides the ability to examine the reconstructed waveform, $r(t)$, obtained in accordance with eq. (1.1) for the case in which

$$s(t) = A_1 \cos(2\pi F_1 t + \theta_1) + A_2 \cos(2\pi F_2 t + \theta_2) \tag{1.3}$$

One property of $s(t)$ and $r(t)$ of interest is whether they are periodic. A waveform, $f(t)$, is periodic if there is a value, T, for which a shift of T seconds results in the same waveform so that

$$f(t) = f(t + T) \tag{1.4}$$

The time T is called a period of $f(t)$. The smallest value of T for which eq. (1.4) is satisfied is called the fundamental period of $f(t)$. Clearly, if T is a period, then so is kT in which k is an integer. As a simple example, consider the sinusoidal waveform

$$f(t) = A \cos [2\pi F t + \phi] \tag{1.5}$$

Then,

$$f(t + T) = A \cos [2\pi F(t + T) + \phi]$$
$$= A \cos [2\pi F t + 2\pi F T + \phi] \tag{1.6}$$

Equation (1.4) is satisfied if $FT = k$, in which k is an integer, since $\cos [\theta] = \cos [\theta + 2\pi k]$. Thus the sinusoid is periodic with periods

$$T = \frac{k}{F} \tag{1.7}$$

in which k is an integer. For $k = 1$, we obtain the fundamental period, which is $1/F$ seconds per cycle.

■ QUESTION 1.3   Let $f(t) = f_1(t) + f_2(t)$ in which $f_1(t)$ is periodic with a fundamental period of T_1 and $f_2(t)$ is periodic with a fundamental period of T_2. Show that $f(t)$ is periodic if and only if

$$\frac{T_1}{T_2} = \frac{n}{m} \tag{1.8}$$

in which n and m are integers.

If T_1 and T_2 satisfy eq. (1.8), they are said to be rationally related because their ratio is a rational number. To obtain the fundamental period of $f(t)$, we first must obtain the smallest values of n and m for which eq. (1.8) is satisfied. This is done by making n and m relatively prime, which means that there are no common integer factors of n and m. For example, 6 and 9 are not relatively prime because 3 is a common factor. However, 2 and 3 are relatively prime. If n and m are relatively prime, then the fundamental period of $f(t)$ is mT_1 (or nT_2).

■ QUESTION 1.4   For each waveform, determine whether it is periodic and, if so, determine its fundamental period.

1. $f_a(t) = \sin(t) + \sin(2t)$
2. $f_b(t) = \sin(t) + \sin(\sqrt{2}t)$
3. $f_c(t) = \cos(\sqrt{2}t) + \cos(2\sqrt{2}t)$
4. $f_d(t) = \sin(2t) + \cos(2\pi t)$

EXPERIMENTAL PROCEDURE

To begin this experiment, run `expr01` from the MATLAB command window. You will be asked to specify the following parameters in eq. (1.3):

- The two sinusoidal frequencies, F_1 and F_2, in hertz
- The two sinusoidal phases, θ_1 and θ_2, in degrees
- The two sinusoidal amplitudes, A_1 and A_2
- The sampling interval, T_s, and the total display time of the signal in seconds, T_o

The waveforms $s(t)$ and $r(t)$ will be drawn in a MATLAB figure window. To differentiate these two waveforms, each waveform will be drawn in a different color. The time axis will be marked at the sampling instances and the ordinate will be marked every 0.1 unit. Observe the 11 cases listed in Table 1.1 for your report. In each case, choose the sampling interval, T_s, to be 0.01 second.

You should also choose other frequencies, amplitudes, and phases in combination with various sampling rates to gain a better understanding of sampling. Discuss these additional cases in your report.

■ QUESTION 1.5   Theoretically show that $s(t)$ and $r(t)$ in each case are periodic waveforms and determine their fundamental periods.

■ QUESTION 1.6   For each case, show that the reconstructed waveform observed is the one predicted by sampling theory.

TABLE 1.1   Cases for Experiment 1 (sampling interval, T_s, is 0.01 second)

	A_1	A_2	F_1(Hz)	F_2(Hz)	θ_1(deg)	θ_2(deg)
1	1	0	20	—	0	—
2	1	0	50	—	60	—
3	1	0	80	—	0	—
4	1	0	80	—	−90	—
5	1	0	100	—	60	—
6	1	1	20	30	0	0
7	1	1	20	80	0	0
8	1	1	20	80	−90	90
9	1	1	20	80	90	90
10	1	1	20	100	0	0
11	1	1	20	120	0	0

■ QUESTION 1.7   For each case, note that $r(t)$ is equal to $s(t)$ at the time instances $t = nT_s$ for any value of F_s even if $F_s < 2F_{max}$. Why? It may help to consider a careful sketch of $g(t)$.

■ QUESTION 1.8   For case 2 in the table, experimentally observe $r(t)$ for various values of the phase θ_1. Use your observations to determine a general expression for $r(t)$ in terms of θ_1.

■ QUESTION 1.9   For case 9 in the table, why is the observed $r(t)$ obvious from a simple observation of $s(t)$? Also explain the observed $r(t)$ in terms of sampling theory.

AN ILLUSTRATION

The MATLAB script that appears when you run expr01 is shown next for parameter values $T_s = 1$, $F_1 = 1.3$, $A_1 = 1$, $A_2 = 0$, and $T_o = 5$.

```
EXPERIMENT 1

THE CONTINUOUS WAVEFORM TO BE SAMPLED IS

s(t) = A1*cos(2*pi*F1*t + Theta_1) + A2*cos(2*pi*F2*t + Theta_2)

THE WAVEFORM s(t) WILL BE SAMPLED EVERY Ts SECONDS AND
THE RECONSTRUCTED WAVEFORM,  r(t), WILL BE OBTAINED BY
"IDEAL"  INTERPOLATION USING SAMPLE VALUES OF s(t).  A
GRAPH OF THE WAVEFORMS s(t) AND r(t) WILL BE DISPLAYED.
```

```
Sampling interval Ts in seconds = 1
     The frequency F1 in hertz = 1.3
     The frequency F2 in hertz = 0
 The phase Theta_1 in degrees = 0
 The phase Theta_2 in degrees = 0
             The amplitude A1 = 1
             The amplitude A2 = 0


THE WAVEFORMS WILL BE DISPLAYED OVER [0, To] INTERVAL
        Value of To in seconds = 5
```

After all the required data have been entered, the graph of $r(t)$ and $s(t)$ will appear in a MATLAB figure window. An example of a graph obtained with the given values is shown in Figure 1.1. Note that waveforms $s(t)$ and $r(t)$ are in different colors in the MATLAB figure window while the sample values $s(nT_s)$ are shown using white circles. The graph in Figure 1.1 with more cycles is $s(t)$. The ordinate is marked every 0.1 unit and the abscissa is marked at the sampling times, which are 0, 1, 2, 3, 4, and 5 seconds for this illustration. Note that the marks at 0 and T_0 seconds are covered by the left and the right ordinates.

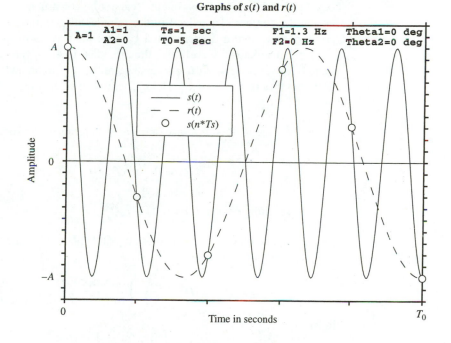

FIGURE 1.1   Graphs of $s(t)$ and $r(t)$ in Experiment 1

Experiment 2

NONIDEAL RECONSTRUCTION

BACKGROUND

As seen in Experiment 1, a bandlimited waveform can be reconstructed from its sample values without error by an ideal D/A converter if the sampling rate is larger than the Nyquist rate. However, the ideal D/A converter described by eqs. (1.1) and (1.2) of Experiment 1 cannot be constructed because it requires an infinite number of past and future samples of $s(t)$ for the interpolation. It could be simulated in Experiment 1 because the waveform sampled, $s(t)$, was known. For unknown waveforms, though, ideal D/A conversion cannot be obtained. However, ideal D/A conversion can be approximated by using only a finite number of terms of eq. (1.1). There are two methods of accomplishing this:

Method A: One method is to use only $(N + 1)$ terms of the sum in eq. (1.1) as

$$r_a(t) = \sum_{n=0}^{N} s(nT_s)g(t - nT_s) \tag{2.1}$$

with the resulting error $e_a(t) = s(t) - r_a(t)$.

Method B: The other method is to use eq. (1.1) but to truncate the interpolation function, $g(t)$, as

$$g_K(t) = g(t)\mathcal{R}\left(\frac{t}{KT_s}\right) = \frac{\sin(\pi F_s t)}{\pi F_s t}\mathcal{R}\left(\frac{t}{KT_s}\right) \tag{2.2}$$

in which

$$\mathcal{R}(\alpha) = \begin{cases} 1 & \text{for } -1 < \alpha < 1 \\ 0 & \text{otherwise} \end{cases} \tag{2.3}$$

is the rectangular function. Note that $g_K(t) = 0$ for $|t| > KT_s$. The resulting error is $e_b(t) = s(t) - r_b(t)$. The error obtained by each of these methods in the interval $0 \le t \le NT_s$ will be examined in this experiment. The objective of studying this approximation error is to obtain a better understanding of ideal D/A conversion. The waveform we shall use for this study is the sinusoid

$$s(t) = \cos(2\pi F t + \theta) \tag{2.4}$$

EXPERIMENTAL PROCEDURE

We begin by experimentally examining the error functions, $e_a(t)$ and $e_b(t)$ in the interval $0 \le t \le NT_s$ obtained using the two methods. To start this experiment, run expr02 from the MATLAB command window. You will be asked to specify the following parameters:

- The sinusoidal frequency, F, in hertz
- The sinusoidal phase, θ, in degrees
- The sampling interval, T_s, in seconds
- The value of N, to be used for Method A
- The value of K, to be used for Method B

Two MATLAB figure windows will be displayed. The first window contains a graph of $s(t)$, $r_a(t)$, and $e_a(t)$; the second window contains a graph of $s(t)$, $r_b(t)$, and $e_b(t)$. You can use the MATLAB zoom function to examine each error function in regions where it is small.

The time required to plot these graphs is proportional to the values of N and K used. Normally, all required observations can be obtained with $3 \le N \le 20$ or $3 \le K \le 20$. Above this value, the separation of the sample points on the graph is somewhat small. However, N and K can be chosen as large as 150 if it is desired to observe the effect of very large values of N or K on the error functions. The time required to obtain graphs is a monotonically increasing function of N or K. This is due to the computational time needed for the evaluation of the interpolation function. To obtain a smooth graph of the reconstructed waveform, it is computed at 20 points per sampling interval and the points are connected by straight lines. With this information, you could experimentally determine the time required for your computer to calculate one value of $r_a(t)$ or $r_b(t)$.

■ QUESTION 2.1   Theoretically, what should be the form of the equation for $T_a(N)$, the computational time in Method A as a function of N, and for $T_b(M)$, the computational time in Method B as a function of M? Explain your reasoning. Is $T_a(0) = 0$? Is $T_b(0) = 0$? Explain. How can you experimentally determine the exact equation for $T_a(N)$ and $T_b(M)$ that is valid for your computer?

To study and compare the error obtained by each of the two methods, first examine the error obtained for the case in which $s(t)$ equals a constant. This is easily obtained by choosing the sinusoidal frequency to be zero. For this study, plot the size of each of the error functions versus N for $N \geq 3$ or K for $K \leq 10$. Of course, you first must determine a meaningful measure of the size of the error functions for this plot. For example, your measure could be the maximum value of $|e(t)|$ in the middle third of the plotted time interval.

■ QUESTION 2.2   Discuss why you believe your measure of the size of the error function is a meaningful one.

Remember that the reconstructed waveform in each method is obtained using only a finite number of samples and that it would be $s(t)$ if N and K were infinite. Thus, the corresponding error function is obtained using the complement of the set of samples used. Use this observation to explain not only the shape of your graph, but also the shape of the error functions. For this, first consider the contribution of the pair of samples immediately to the right and to the left of the samples used. Then add the contribution of the next pair of right and left samples, and so on.

■ QUESTION 2.3   In what manner does the determination of the reconstructed waveform obtained by Method A differ from that of Method B? Illustrate the difference between the two determinations graphically.

■ QUESTION 2.4   Compare the observed error functions $e_a(t)$ and $e_b(t)$. As in ideal D/A conversions, is $e_a(nT_s) = 0$? Is $e_b(nT_s) = 0$?

To further study the error functions, you first should experimentally determine whether the error is invariant with any function of the parameters. For example, you have seen in ideal D/A conversion (in which N and K are infinite) that the error is an invariant function of the relative frequency, $f = F/F_s$, in which F is the sinusoidal frequency and F_s is the sampling rate.

■ QUESTION 2.5   Is $e_a(t)$ and/or $e_b(t)$ also dependent only on the relative frequency f? Verify your conclusions theoretically.

For $N = 4$ and 5 and for $K = 4$ and 5, plot the size of each error function versus θ using your measure of size for $f = 0.15, 0.30$, and 0.45. Compare your graphs and discuss your results.

AN ILLUSTRATION

The MATLAB script that appears when you run expr2 is shown below for the parameter values $F = 1$, $\theta = 0$, $T_s = 0.1$, $N = 10$, and $K = 5$.

```
                    EXPERIMENT 2

         THE CONTINUOUS WAVEFORM TO BE SAMPLED IS


             s(t) = cos(2*pi*F*t + Theta)


THE WAVEFORM s(t) WILL BE SAMPLED EVERY Ts SECONDS AT THE TIMES
t = n*Ts FOR n = 0, 1, ...,N. THE RECONSTRUCTED WAVEFORMS, ra(t)
AND rb(t), WILL BE OBTAINED USING TWO NONIDEAL METHODS. METHOD A
USES ONLY THE N SAMPLE VALUES OF s(t) WHILE METHOD B USES A
K-TH-ORDER TRUNCATED SINC INTERPOLATION FUNCTION.
GRAPHS OF THE WAVEFORMS:
     1. s(n*Ts) & r(t); ra(t) & ea(t) = r(t) - ra(t),
     2. s(n*Ts) & r(t); rb(t) & eb(t) = r(t) - rb(t).
WILL BE AVAILABLE FOR DISPLAY IN TWO FIGURE WINDOWS.


         Sinusoidal frequency F in hertz = 1
         Sinusoidal phase Theta in degrees = 0
         Sampling interval Ts in seconds = 0.1
The value of N in Method A (3 <= N <= 20)  = 10
The value of K in Method B (3 <= K <= 20)  = 5
```

After you enter all the required data, the two MATLAB figure windows described above will appear. An example of each window for the given parameter values is shown in Figures 2.1 and 2.2. Two waveforms are differentiated by different colors on the same graph while the sample values are shown using white circles. The graph ordinate is marked every 0.1 unit and the abscissa is marked at the sampling times.

■ QUESTION 2.6 Why is $e(t) = 0$ at the sampling times?

■ QUESTION 2.7 Discuss why the smallest values of the maxima of $e(t)$ occur at about the center of the interval.

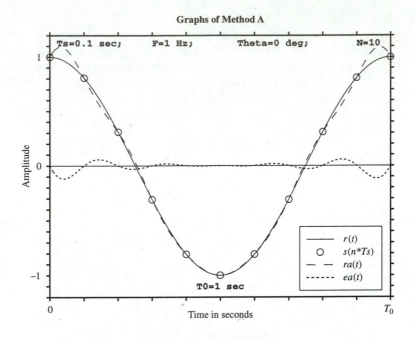

FIGURE 2.1   Plots using Method A in Experiment 2

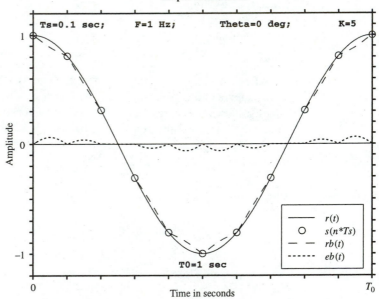

FIGURE 2.2   Plots using Method B in Experiment 2

Experiment 3

PRACTICAL INTERPOLATORS

BACKGROUND

If the sampling rate is larger than the Nyquist rate, then, as seen in Experiment 2, a bandlimited waveform can be reconstructed from its sample values with an acceptable error by using a truncated version of the ideal D/A converter. It should be noted that, for a reasonably small error, the techniques discussed in Experiment 2 require a large amount of computation to determine $r(t)$. For this reason, D/A conversion is usually accomplished using other choices for the interpolation function that are less computationally intensive. A brief discussion of practical D/A conversion is presented here since it is not included in most texts.

For the input sample values, $s(nT_s)$, the output, $r(t)$, of a linear, time-invariant (LTI) D/A converter can be expressed as the convolutional sum

$$r(t) = \sum_{n=-\infty}^{\infty} s(nT_s)v(t - nT_s) \tag{3.1}$$

■ QUESTION 3.1 It is desirable that a D/S converter be a linear system even if it is not time-invariant. Discuss why a D/A converter should be a linear system.

Ideal D/A conversion is obtained if in eq. (3.1) the interpolation function, $v(t)$, is the ideal one, $g(t)$, given by eq. (1.2). With the ideal function, $g(t)$, we observed in Experiment 1 that $r(nT_s) = s(nT_s)$ even if there is aliasing. We noted that this is a consequence of the fact that $g(0) = 1$ and $g(nT_s) = 0$ for $n = \pm 1, \pm 2, \pm 3, \ldots$. However, the ideal D/A converter is not causal because $g(t) \neq 0$ for $t < 0$ so that future values of $s(nT_s)$ are required to form $r(t)$. Also, since $g(t)$ only approaches zero as $1/t$, all past values of $s(nT_s)$ are required to form $r(t)$ and so must be stored.

To design a causal LTI D/A converter that does not require infinite storage, first consider a D/A converter with an interpolation function $g_k(t)$ described in Experiment 2, eq. (2.2). Since this function is zero for $|t| > kT_s$, the reconstructed waveform is obtained using k past values of $s(nT_s)$, the present value of $s(nT_s)$, and k future values of $s(nT_s)$ for a total of $(2k + 1)$ sample values of $s(t)$. Such a D/A converter still would not be causal because $g_k(t) \neq 0$ for $t < 0$. However, it can be made causal by using a delay of kT_s seconds. This is easily accomplished by using $g_k(t - kT_s)$ as the interpolation function.

From this discussion, we note that, for a practical LTI D/A converter, there are three desirable properties the interpolation function, $v(t)$, should satisfy:

1. $v(0) = 1$
2. $v(nT_s) = 0$ for $n = \pm1, \pm2, \pm3, \ldots$
3. $v(t) = 0$ for $|t| > kT_s$, for some integer k

■ QUESTION 3.2   Carefully explain why these three conditions are desirable properties.

In this experiment, the resulting error, $e(t) = s(t) - r(t)$, will be examined for some different practical D/A converters with the objective of obtaining a better understanding of D/A conversion.

PART 1: K^{TH}-ORDER POLYNOMIAL INTERPOLATION

───────────────────■───────────────────

One class of choices for a practical interpolation function, $v(t)$, is the class of kth-order polynomial interpolation functions. A kth-order polynomial D/A reconstruction is obtained by connecting the values of successive samples by a kth-degree polynomial.

Zero-Order Polynomial Interpolation   This reconstruction is a piecewise constant curve in which the constant is equal to the last sample value. This value is held until the next sample value is received. The reconstructed waveform is a staircase approximation of $s(t)$.   •

■ QUESTION 3.3   Show that zero-order reconstruction is obtained using the interpolation function

$$h_0(t) = \begin{cases} 1 & \text{for } 0 \leq t \leq T_s \\ 0 & \text{otherwise} \end{cases} \qquad (3.2)$$

which is shown in Figure 3.1.

First-Order Polynomial Interpolation This reconstruction is obtained by connecting adjacent sample values by a first-degree curve, $a + bt$.

■ QUESTION 3.4 Using the three conditions that an interpolation function, $v(t)$, should satisfy, show that first-order reconstruction is obtained by using the interpolation function

$$h_{01}(t) = \left[1 - \frac{|t|}{T_s}\right] \mathcal{R}\left(\frac{t}{T_s}\right) \tag{3.3}$$

in which

$$\mathcal{R}(\alpha) = \begin{cases} 1 & \text{for } -1 < \alpha < 1 \\ 0 & \text{otherwise} \end{cases} \tag{3.4}$$

is the rectangular function.

The interpolation function $h_{01}(t)$ is shown in Figure 3.1. For causality, the D/A converter would use $h_{01}(t - T_s)$ as its interpolation function, $v(t)$. Physically, the delay of T_s seconds is required to determine the slope of the straight line that connects the present sample value with the next one.

Second-Order Polynomial Interpolation In this reconstruction, adjacent sample values are connected by a second-degree curve, $a + bt + ct^2$, in which the constants a, b, and c are chosen so that the reconstructed curve passes through the sample values at the interval endpoints.

■ QUESTION 3.5 Using the three conditions that an interpolation function, $v(t)$, should satisfy, show that second-order reconstruction is obtained by using the interpolation function

$$h_{02}(t) = \frac{1}{2}\left[2 - 3\frac{|t|}{T_s} + \left(\frac{|t|}{T_s}\right)^2\right] \mathcal{R}\left(\frac{t}{2T_s}\right) \tag{3.5}$$

The interpolation function $h_{02}(t)$ is shown in Figure 3.1. For causality, the D/A converter would use $h_{02}(t - 2T_s)$ as its interpolation function, $v(t)$. Note that a second-order reconstruction requires a delay of $2T_s$ seconds.

Higher-Order Polynomial Interpolation In a similar fashion, we can extend the approach of reconstruction to higher-order interpolations.

■ QUESTION 3.6 Show that a causal third-order polynomial D/A converter would use $h_{03}(t - 3T_s)$ as its interpolation function, $v(t)$, in which

$$h_{03}(t) = \frac{1}{6}\left[6 - 11\frac{|t|}{T_s} + 6\left(\frac{|t|}{T_s}\right)^2 - \left(\frac{|t|}{T_s}\right)^3\right] \mathcal{R}\left(\frac{t}{3T_s}\right) \tag{3.6}$$

which is shown in Figure 3.1. Remember to use the three conditions that an interpolation function, $v(t)$, should satisfy.

Note that a third-order reconstruction requires a delay of $3T_s$ seconds. In general, a kth-order reconstruction requires a delay of kT_s seconds. Depending on the waveform, $s(t)$, the errors obtained using a kth-order reconstruction can be greater or smaller than the approximation error, $e(t)$, obtained in Experiment 2. A zero-order polynomial is used when it is desired to minimize the amount of computation required to obtain the reconstructed waveform. A first-order polynomial is used when it is desired that the reconstructed waveform be a continuous curve. The continuous waveforms displayed on your monitor with the experiments of this text are obtained by connecting the computed values with straight lines. Thus, they are obtained using first-order D/A conversion.

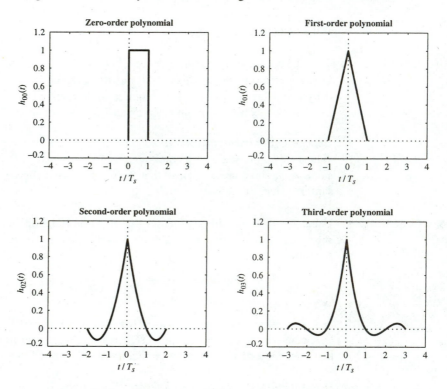

FIGURE 3.1   The kth-order polynomial interpolation functions

EXPERIMENTAL PROCEDURE   For $s(t) = \cos(2\pi F t + \theta)$, compare the reconstructed waveform $r(t)$ obtained using zero-, first-, second-, and third-order polynomial reconstruction for various frequencies F and phases θ. Which order polynomial results in the more desirable reconstructed waveforms? For this, first formulate a set of desirability criteria. Discuss your choice of desirability criteria and why you believe your criteria are good ones. Compare the reconstructed waveforms in terms of your criteria. Discuss the properties of each interpolation function that make the resulting reconstructed waveform more (or less) desirable.

PART 2: K^{TH}-ORDER POLYNOMIAL DIFFERENTIABLE INTERPOLATION .

Some applications require that the derivative of the reconstructed waveform be a continuous curve. For such applications, the kth-order polynomial reconstruction described in Part 1 cannot be used since the derivative of the reconstructed waveform is discontinuous at the times $t = nT_s$ for $n = 0$ and $\pm kT_s$.

■ QUESTION 3.7   Show that the discontinuity of the derivative of the reconstructed waveform, $r(t)$ at the times $t = nT_s$, obtained by k^{th}-order polynomial reconstruction for $k \geq 1$ is

$$D_k(nT_s) = \left\{ h'_{0k}(0^+) - h'_{0k}(0^-) \right\} s[nT_s] + h'_{0k}(-kT_s^+)s[(n+k)T_s]$$
$$-h'_{0k}(-kT_s^-)s[(n-k)T_s] \qquad (3.7)$$

in which the prime indicates the derivative of the function. Verify this result for the case of first-order polynomial reconstruction by comparing $D_1(nT_s)$ with the change of slope of the straight lines at the sample value $s(nT_s)$.

The derivative of the reconstructed waveform would be continuous if the derivative of the interpolation function, $v(t) = h_{0k}(t - kT_s)$, were continuous. Note from eq. (3.7) that this condition can be satisfied by requiring the polynomial interpolation function to also satisfy the condition[1]

4a. $v'(t + kT_s) = 0$ for $t = 0$, $k\,T_s^-$, and $-k\,T_s^+$

This condition is in addition to the three earlier conditions that $v(t)$ should satisfy. Note that we could make the first m derivatives of the reconstructed waveform, $r(t)$, continuous by imposing the condition that the first m derivatives of the interpolation function, $v(t)$, be continuous. Thus, the first m derivatives of $r(t)$ are continuous if we replace condition (4a) with the condition that for $p = 1, 2, ..., m$,

4b. $v^{(p)}(t + kT_s) = 0$ for $t = 0, k\,T_s^-$, and $-k\,T_s^+$

---

[1] From eq. (3.7), it is only necessary that $h'_{0k}(0^-) = h'_{0k}(0^+) = x$. Note that if $x = x_0$ results in the best reconstructed waveform for the sample values of some sequence, then $x = -x_0$ would result in the best reconstructed waveform for the negative of that sequence. We thus have chosen $x = 0$ for this experiment.

We shall use the notation $h_{mk}(t)$ to indicate a polynomial interpolation function that is zero for $|t| \geq kT_s$ and for which its first m derivatives are continuous. Functions for which the first m derivatives are continuous are called functions of the class C^m. Note that $m = 0$ for the interpolation functions discussed in Part 1.

The polynomial interpolation function of class C^1, which is zero for $|t| > T_s$, is

$$h_{11}(t) = \left[1 - 3\left(\frac{t}{T_s}\right)^2 + 2\left(\frac{|t|}{T_s}\right)^3 \right] \mathcal{R}\left(\frac{t}{T_s}\right) \qquad (3.8)$$

and the polynomial interpolation function of class C^2, which also is zero for $|t| > T_s$, is

$$h_{21}(t) = \left[1 - 10\left(\frac{|t|}{T_s}\right)^3 + 15\left(\frac{|t|}{T_s}\right)^4 - 6\left(\frac{|t|}{T_s}\right)^5 \right] \mathcal{R}\left(\frac{t}{T_s}\right) \qquad (3.9)$$

These functions are shown in Figure 3.2.

■ QUESTION 3.8   Derive the expression for $h_{11}(t)$ using the four conditions given above.

For causality, the D/A converter would use $h_{m1}(t - T_s)$ as its interpolation function, $v(t)$, so that a delay of only one sample is required.

EXPERIMENTAL PROCEDURE   For $s(t) = \cos(2\pi Ft + \theta)$, compare the reconstructed waveform $r(t)$ obtained using the interpolation functions $h_{01}(t)$, $h_{11}(t)$, and $h_{21}(t)$ for various frequencies F and phases θ. Which interpolation function results in the more desirable reconstructed waveform? For this, compare the reconstructed waveforms in terms of the desirability criteria you used in Part 1. Discuss the properties of each interpolation function that make the resulting reconstructed waveform more (or less) desirable.

■ QUESTION 3.9   Show that $\lim_{m \to \infty} h_{m1}(t) = h_0(t + \frac{1}{2}T_s)$ in which $h_0(t)$ is the zero-order polynomial interpolation function. In accordance with this result, we can define $h_{00}(t) = h_0(t + \frac{1}{2}T_s)$ as the 0th member of the $h_{0k}(t)$ family.

We now examine the reconstructed waveform $r(t)$ obtained using the polynomial interpolation functions $h_{m2}(t)$, which are zero for $|t| \geq 2T_s$. Using the four conditions stated above, we obtain

$$h_{12}(t) = \frac{1}{4}\left[4 - 11\left(\frac{|t|}{T_s}\right)^2 + 9\left(\frac{|t|}{T_s}\right)^3 - 2\left(\frac{|t|}{T_s}\right)^4 \right] \mathcal{R}\left(\frac{t}{2T_s}\right) \qquad (3.10)$$

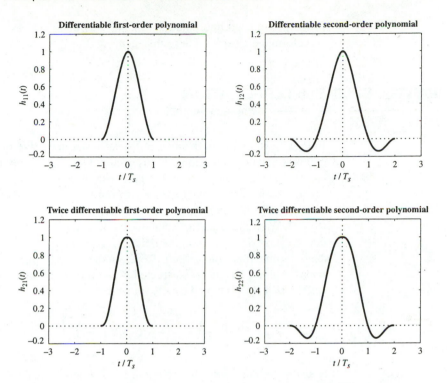

FIGURE 3.2   The kth-order polynomial differentiable interpolation functions

and

$$h_{22}(t) = \frac{1}{16} \left[16 - 84 \left(\frac{|t|}{T_s} \right)^3 + 111 \left(\frac{|t|}{T_s} \right)^4 - 51 \left(\frac{|t|}{T_s} \right)^5 + 8 \left(\frac{|t|}{T_s} \right)^6 \right] \times$$

$$\mathcal{R} \left(\frac{t}{2T_s} \right) \tag{3.11}$$

These functions are shown in Figure 3.2.

For $s(t) = \cos(2\pi F t + \theta)$, compare the reconstructed waveform $r(t)$ obtained using the interpolation functions $h_{m2}(t)$ for $m = 0$, 1, and 2 for various frequencies F and phases θ. Which interpolation function results in the more desirable reconstructed waveform? For this, compare the reconstructed waveforms in terms of the desirability criteria you used in Part 1. Discuss the properties of each interpolation function that make the resulting reconstructed waveform more (or less) desirable.

Compare the results obtained using the polynomial interpolation functions for $h_{m1}(t)$ with those obtained for $h_{m2}(t)$. For this, compare the reconstructed waveforms in terms of the desirability criteria you used in Part 1. Discuss the properties of each interpolation function that make the resulting reconstructed waveform more (or less) desirable.

PART 3: SPLINE INTERPOLATION

In Part 2, the first m derivatives of the reconstructed waveform $r(t)$ were made continuous using condition (4b). You should have observed that this could lead to $r(t)$ having some undesirable properties. Another practical approach is to first determine the reconstructed waveform $r(t)$ directly in terms of the sample values $s(nT_s)$ and then determine the interpolation function from the expression for $r(t)$. We'll illustrate the technique by determining a third-degree polynomial interpolation such that $r(t)$ is of class C^1. Such a polynomial is called a third-degree spline (or cubic spline). To determine a cubic spline interpolation of the sample values, let the reconstructed waveform in the interval between the nth and the $(n + 1)^{\text{st}}$ sample be $r_n(t)$, in which

$$r_n(t) = \left[1 + a\left(\frac{t}{T_s}\right) + b\left(\frac{t}{T_s}\right)^2 + c\left(\frac{t}{T_s}\right)^3 \right] s(nT_s), \quad \text{for } 0 \leq \frac{t}{T_s} \leq 1$$

(3.12)

Note that the constant in the polynomial was chosen equal to one so that $r_n(0) = s(nT_s)$. The other three constants, a, b, and c, are determined by using these conditions:

S1. $r_n(T_s) = s\left[(n + 1)T_s\right]$

S2. $r_n'(0) = \dfrac{s\left[(n + 1)T_s\right] - s\left[(n - 1)T_s\right]}{2T_s}$

S3. $r_n'(T_s) = \dfrac{s\left[(n + 2)T_s\right] - s\left[nT_s\right]}{2T_s}$

The first condition together with $r_n(0) = s(nT_s)$ makes $r(nT_s) = s(nT_s)$. The second and third conditions make the derivative of $r(t)$ at $t = nT_s$ continuous and equal to the slope of a straight line connecting the sample values to the right and the left of $s(nT_s)$. This slope is equal to the average of the slope of a straight line connecting $s(nT_s)$ with $s[(n - 1)T_s]$ and the slope of a straight line connecting $s(nT_s)$ with $s[(n + 1)T_s]$.

■ **QUESTION 3.10**   Use the three conditions to show that the constants a, b, and c in eq. (3.12) are

$$a = \frac{s\left[(n-1)T_s\right] - s\left[(n-1)T_s\right]}{2s(nT_s)}$$

$$b = \frac{-s\left[(n+2)T_s\right] + 4s\left[(n+1)T_s\right] - 5s(nT_s) + 2s\left[(n-1)T_s\right]}{2s(nT_s)}$$

$$c = \frac{s\left[(n+2)T_s\right] - 3s\left[(n+1)T_s\right] + 3s(nT_s) - s\left[(n-1)T_s\right]}{2s(nT_s)}$$

$$(3.13)$$

We observe that, for any given value of t, eq. (3.12) is a linear combination of the sample values $s[(n-1)T_s]$, $s[nT_s]$, $s[(n+1)T_s]$, and $s[(n+2)T_s]$. We thus should be able to obtain $r(t)$ as the response of a LTI system since the output of a LTI system is a linear combination of past, present, and future (if the system is noncausal) values of its input. To determine the unit impulse response of the system, we use eqs (3.12) and (3.13) to express $r_n(t)$ as

$$r_n(t) = \left[-\frac{1}{2}\left(\frac{t}{T_s}\right) + \left(\frac{t}{T_s}\right)^2 - \frac{1}{2}\left(\frac{t}{T_s}\right)^3 \right] s\left[(n-1)T_s\right]$$

$$+ \left[1 - \frac{5}{2}\left(\frac{t}{T_s}\right)^2 + \frac{3}{2}\left(\frac{t}{T_s}\right)^3 \right] s\left[nT_s\right]$$

$$+ \left[\frac{1}{2}\left(\frac{t}{T_s}\right) + 2\left(\frac{t}{T_s}\right)^2 - \frac{3}{2}\left(\frac{t}{T_s}\right)^3 \right] s\left[(n+1)T_s\right]$$

$$+ \left[-\frac{1}{2}\left(\frac{t}{T_s}\right)^2 + \frac{1}{2}\left(\frac{t}{T_s}\right)^3 \right] s\left[(n+2)T_s\right],$$

$$\text{for } 0 \le \left(\frac{t}{T_s}\right) \le 1 \qquad\qquad (3.14)$$

■ QUESTION 3.11   Show that if the unit sample response of the LTI D/A converter were

$$h_s(t) = \begin{cases} g_a(t) & \text{for } -2T_s < t < -T_s \\ g_b(t) & \text{for } -T_s < t < 0 \\ g_c(t) & \text{for } 0 < t < T_s \\ g_d(t) & \text{for } T_s < t < 2T_s \\ 0 & \text{otherwise} \end{cases} \qquad (3.15)$$

then for the input sample values, the output of the D/A converter for $nT_s < t < (n+1)T_s$ would be

$$r(t) = g_a\left[t - (n+2)T_s\right] s\left[(n+2)T_s\right] + g_b\left[t - (n+1)T_s\right] s\left[(n+1)T_s\right]$$

$$+ g_c\left[t - nT_s\right] s\left[nT_s\right] + g_d\left[t - (n-1)T_s\right] s\left[(n-1)T_s\right] \qquad (3.16)$$

By comparing eqs. (3.16) with (3.14), we note that they are equal if

$$g_a(t - 2T_s) = -\frac{1}{2}\left(\frac{t}{T_s}\right)^2 + \frac{1}{2}\left(\frac{t}{T_s}\right)^3 \tag{3.17a}$$

$$g_b(t - T_s) = \frac{1}{2}\left(\frac{t}{T_s}\right) + 2\left(\frac{t}{T_s}\right)^2 - \frac{3}{2}\left(\frac{t}{T_s}\right)^3 \tag{3.17b}$$

$$g_c(t) = 1 - \frac{5}{2}\left(\frac{t}{T_s}\right)^2 + \frac{3}{2}\left(\frac{t}{T_s}\right)^3 \tag{3.17c}$$

$$g_d(t + T_s) = -\frac{1}{2}\left(\frac{t}{T_s}\right) + \left(\frac{t}{T_s}\right)^2 - \frac{1}{2}\left(\frac{t}{T_s}\right)^3 \tag{3.17d}$$

The cubic spline interpolation function, $h_s(t)$, given by eqs. (3.15) and (3.17) is shown in Figure 3.3, which also shows the function $h_{12}(t)$ for comparison. For causality, the LTI D/A converter would use $h_s(t - 2T_s)$ as its interpolation function, $v(t)$. Note that the cubic spline interpolation requires a delay of $2T_s$ seconds.

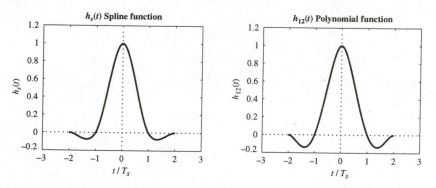

FIGURE 3.3   (a) The cubic spline interpolation function, $h_s(t)$; (b) the polynomial function, $h_{12}(t)$

■ QUESTION 3.12   Show that the cubic spline interpolation function, $h_s(t)$, is of class C^1 so that the reconstructed waveform, $r(t)$, also is of class C^1.

EXPERIMENTAL   Examine the cubic spline interpolation for various values of F and θ and, us-
PROCEDURE   ing your desirability criteria, compare the reconstructed waveform with those obtained in Parts 1 and 2. Discuss the specific differences between $h_{12}(t)$ and $h_s(t)$. Also discuss the specific differences between the reconstructed waveform $r(t)$ obtained using $h_{12}(t)$ and that obtained using $h_s(t)$.

AN ILLUSTRATION

This experiment is available in the function `expr3`. When you run `expr3`, the following MATLAB script appears.

```
EXPERIMENT 3: STUDY OF PRACTICAL INTERPOLATION TECHNIQUES

    THE CONTINUOUS WAVEFORM TO BE SAMPLED IS
            s(t) = cos(2*pi*F*t + Theta)
 THE WAVEFORM s(t) WILL BE SAMPLED EVERY Ts SECONDS AT THE TIMES
t = k*Ts FOR k = 0, 1, ...,N-1 AND THE RECONSTRUCTED WAVEFORM,
r(t), WILL BE OBTAINED FROM THE N SAMPLE VALUES OF s(t) USING
"PRACTICAL" INTERPOLATION FUNCTIONS.  THESE FUNCTIONS ARE
GROUPED IN THREE PARTS:

  PART A: CONTINUOUS kTH-ORDER HOLD FUNCTIONS, k = 0,1,2,3
  PART B: DIFFERENTIABLE kTH-ORDER HOLD FUNCTIONS, k = 1,2
  PART C: CUBIC SPLINE FUNCTION

PRESS ANY KEY TO CONTINUE.
```

The experiment has three parts as we discussed. When you press any key, a MATLAB figure window, shown next, appears that provides choices for the three parts and an option to end the experiment.

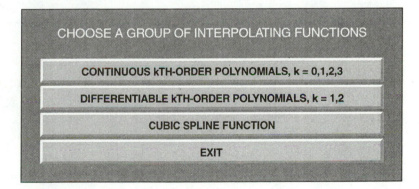

The MATLAB script that appears when you choose Part 1 is shown next for the parameter values $F = 1$, $\theta = 0$, $F_s = 10$, and $N = 11$.

```
PART 1: CONTINUOUS kTH-ORDER HOLD FUNCTIONS, k = 0,1,2,3

  THE CONTINUOUS WAVEFORM TO BE SAMPLED IS
```

```
s(t) = cos(2*pi*F*t + Theta); t = k/Fs FOR k = 0, 1, ...,N-1
```

```
           Frequency, F, in hertz = 1
          Phase, Theta, in degrees = 0
Sampling rate, Fs, in samples/second = 10
  Number of samples (3 < N <= 20), N = 11
```

The other two parts have similar scripts. After you enter all the required data, a MATLAB figure window appears, as shown in Figure 3.4 for the given parameter values. There are four subplots in which the original and the interpolated waveforms using different interpolation functions described in Part 1 are shown. Two waveforms are shown using different colors on the same graph. The abscissa is marked at the sampling times. Similar MATLAB figure windows appear in Parts 2 and 3 and are shown in Figures 3.5 and 3.6 respectively.

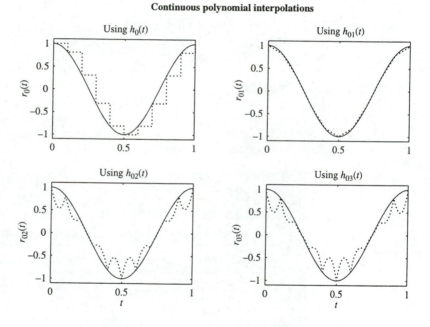

FIGURE 3.4   MATLAB plots in Part 1 of Experiment 3

Differentiable polynomial interpolations

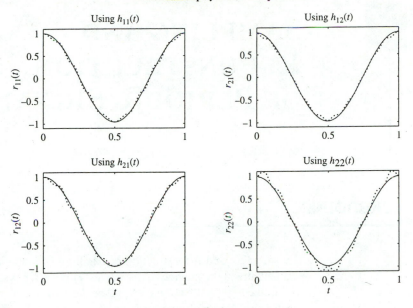

FIGURE 3.5 MATLAB plots in Part 2 of Experiment 3

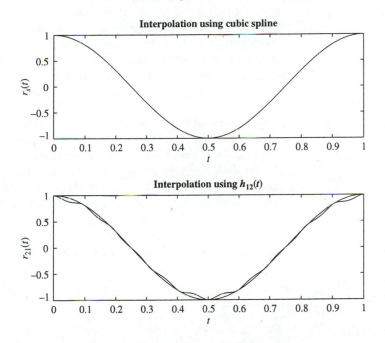

FIGURE 3.6 MATLAB plots in Part 3 of Experiment 3

Experiment 4

SAMPLING AND RECONSTRUCTION OF PERIODIC SIGNALS

BACKGROUND

The sampling and reconstruction of bandlimited waveforms were studied experimentally in the previous experiments. However, physical waveforms are not bandlimited since it can be shown that the future of a bandlimited waveform can be predicted with arbitrarily small error from a short-time sample of the waveform.[1] Consequently, no information-bearing signal is bandlimited since, if it were bandlimited, no new information could be inserted at any time in the future. We thus conclude that no interesting waveform is bandlimited. However, for certain analyses, many such waveforms can be approximated as a bandlimited waveform without much violence to the final result. The error obtained due to a bandlimited waveform approximation should always be checked to be certain that the resulting error incurred in the analysis is within an acceptable bound.

Since physical waveforms are not bandlimited, there always will be aliasing no matter what the sampling rate. One method used to reduce this aliasing error is to place a low-pass filter before the sampler as shown in Figure 4.1. Such a filter is called an anti-aliasing filter. For a given sampling rate, the cut-off frequency of the filter is chosen to reduce the aliasing error. The filter will, of course, distort the waveform. The higher the filter cut-off frequency, the less is this distortion. However, increasing the filter cut-off frequency increases the amount of aliasing. The filter cut-off frequency is chosen so that the total error due to the distortion and due to aliasing is a minimum.

---

[1] This result is due to the mathematician N. Wiener. A development of his result is contained in his classic text, *Extrapolation, Interpolation, and Smoothing of Stationary Time Series with Engineering Applications*, John Wiley & Sons, Inc., New York, NY, 1949. Also see M. Schetzen and A. A. Al-Shalchi, "Prediction of Singular Time Functions," Quarterly Progress Report No. 67, *Research*

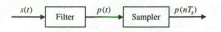

FIGURE 4.1 Anti-aliasing filter

To experimentally study this method, the system shown in Figure 4.2 is implemented by the computer program. The implemented filter is an ideal low-pass filter. That is, the filter will pass all sinusoids with frequencies less than and equal to the cut-off frequency without a change in the sinusoidal amplitude or phase and will not pass sinusoids with frequencies larger than the cut-off frequency. The sampler and the ideal D/A converter are the same as in Experiment 1.

FIGURE 4.2 System for Experiment 4

The signal, $s(t)$, used for this study is the periodic trapezoidal waveform shown in Figure 4.3. As shown, the fundamental period is T and the delay is the time from the center of the interval during which $s(t) = 1$. The sloping section is determined in the computer program by specifying d/T; the waveform is a square wave if $d/T = 0$ and it is a triangular waveform if $d/T = 0.50$. Values of d/T that are negative or larger than 0.50 cannot be chosen. The waveform is sampled at the times $t = nT_s$ in which T_s is the sampling interval. The point in the period at which the first sample is taken thus is determined by specifying the delay. A positive or negative delay can be chosen; a positive delay is illustrated in Figure 4.3.

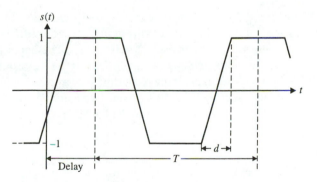

FIGURE 4.3 Trapezoidal waveform used in Experiment 4

Laboratory of Electronics, M.I.T., Cambridge, MA, October 15, 1962, pp. 125–137.

EXPERIMENTAL PROCEDURE

■

Run `expr04` from the MATLAB command window to begin this experiment. You will be asked to specify the following parameters given in Figure 4.3:

- The fundamental period, T, in seconds
- The slope, d/T
- The delay in seconds
- The filter cut-off frequency in hertz
- The sampling interval, T_s, in seconds

Two MATLAB figure windows will be displayed. The first window contains three graphs, one each for $s(t)$, $p(t)$, and $r(t)$ waveforms shown in Figure 4.2. The second window contains three graphs, each showing a pair of waveforms (in different colors) for comparison purposes.

PART 1

To better understand the importance of the sampling rate on the reconstructed waveform and also to become acquainted with the program, first experiment by choosing the parameters of the waveform, $s(t)$, to be $d/T = 0$ and $T = 1$ second. This will result in $s(t)$ being a square-wave with a fundamental period of 1 second. Choose the filter cut-off frequency to be 150 Hz,[2] delay equal to 0, the sampling frequency F_s to be 6.5 samples per second, and observe $r(t)$, the output of the ideal D/A converter.

■ QUESTION 4.1   What is the fundamental period of $r(t)$?

Now choose other values of T_s such as 0.14, 0.15, 0.17, and 0.25 second. Compare the various waveforms $r(t)$ obtained. For example, compare the fundamental period of each and also other waveshape characteristics. Thus, note how the sampling rate can significantly affect the resulting reconstructed waveform.

The graph of the reconstructed waveform, $r(t)$, can be understood better by considering it in the frequency domain. For this, first note from Experiment 1 that the sampler and also the D/A converter are linear systems. Thus, if $s(t)$ is

---

[2]The maximum cut-off frequency that can be chosen has been set to be the 150th harmonic of $s(t)$ since, as a result of computer and screen quantization, the error reduction obtained by using more harmonics is not observable on the screen. Limiting the highest cut-off frequency to be the 150th harmonic of $s(t)$ thus does not reduce the quality of any of the observed graphs, while it does limit the amount of computation (and thus the time) required to produce some of the graphs. If you choose a cut-off frequency larger than the 150th harmonic, the program automatically will change it to the 150th harmonic and inform you of the change. Because the filter cut-off was chosen to be 150 Hz for this experiment, the filter output, $p(t)$, will appear identical to $s(t)$.

expressed as the sum of a number of individual waveforms as

$$s(t) = \sum_n s_n(t) \tag{4.1}$$

then $r(t)$ can be expressed as

$$r(t) = \sum_n r_n(t) \tag{4.2}$$

in which $r_n(t)$ is the reconstructed waveform due to $s_n(t)$. Since we know the ideal D/A output when the sampler input is a sinusoid, we choose to represent $s(t)$ as the sum of sinusoids. This, of course, is the Fourier series of $s(t)$.

■ QUESTION 4.2 Obtain the Fourier series of $s(t)$ in terms of the parameters d, T, and delay in the form

$$s(t) = \sum_{n=0}^{\infty} a_n \cos(\omega_n t + \varphi_n) \tag{4.3}$$

Your expression can be verified by observing $p(t)$ for various choices of the filter cut-off frequency, d/T, and delay. For example, use $T = 1$ second and $T_s = 0.01$ second, and set the cut-off frequency so that the filter passes only the first harmonic of $s(t)$. Then, $p(t)$ is a sinusoid. Is its amplitude and phase that which you calculated? Now, compare your calculation with each $p(t)$ obtained with the filter cut-off chosen to pass the second, third, and fourth harmonics. Observe the dependence of the rate of convergence on d/T. What is the dependence of a_n on n for $d/T = 0$ and for $d/T = 0.5$?

The reconstructed waveform, $r(t)$, now can be obtained by determining the alias of each of the Fourier components of $s(t)$. Compare your calculation with the observed $r(t)$ for each of the following cases with $d/T = 0$, $T = 1$, $T_s = 0.50$, delay $= 0$, and filter cut-off frequency $= 2, 4$, and 6 Hz. Repeat with delay $= 0.25$ second. Now, choose the parameters to be $d/T = 0$, $T = 1$, $T_s = 0.1$, delay $= 0.25$, and filter cut-off frequency $= 6$ Hz. Explain the observed $r(t)$. For this, it might be helpful to review Experiment 1.

PART 2 The optimum filter cut-off frequency will be determined in this part of the experiment. First choose the parameters to be $d/T = 0$, $T = 1$, $T_s = \frac{1}{12}$, and delay $= 0$. Now compare $s(t)$, $p(t)$, and $r(t)$ for various choices of the filter cut-off frequency. What is the optimum choice of the cut-off frequency? For this, you first need to define what you mean by *optimum*. This is similar to the problem of choosing a measure of the error in Experiment 2. Different definitions of *optimum* can result in different conclusions. Many problems in science and engineering first require a definition of *optimum* or *best* that is meaningful for the given problem. This is not always a simple matter since the definition must not only be meaningful for the given problem but also lead to a reasonable theoretical analysis. Choose a definition of *optimum* and discuss why you believe your choice is meaningful for this experiment. Determine the optimum value of the cut-off frequency using your definition of *optimum*. Discuss your result.

■ QUESTION 4.3   Is your value of the optimum cut-off frequency affected by the value of the delay? Discuss your result.

Change d/T to 0.5 so that $s(t)$ is a triangular waveform and use the same parameter values as above to determine the optimum cut-off frequency. Is it the same as that obtained previously with the square wave? Is the optimum filter cut-off frequency what you originally expected? Now, experimentally determine the optimum cut-off frequency for other values of d/T between 0 and 0.5. Discuss your results.

AN ILLUSTRATION

The MATLAB script that appears when you run expr04 is shown below for the parameter values $T = 1$, $d/T = 0.1$, delay $= 0$, cut-off frequency $= 4$, and $T_s = 1$. As discussed above, the maximum cut-off frequency that can be chosen is equal to the 150th harmonic of $s(t)$.

```
                     EXPERIMENT 4

        ***  PARAMETERS OF THE PERIODIC WAVEFORM s(t)  ***
   Fundamental period T1 in seconds = 1
                      Slope d/T = 0.1


        ***  SAMPLING PARAMETERS  ***
              Delay in seconds  = 0
 Prefilter cutoff frequency in hertz = 4
    Sampling interval Ts in seconds = 1
```

Individual waveforms $s(t)$, $p(t)$, and $r(t)$ are shown in Figure 4.4. Figure 4.5 shows two waveforms in one graph. The abscissa is marked at the sampling instances and the ordinate is marked every 0.1 unit. Should $r(t) = s(t)$ at the sampling instances, as was the case in the previous experiments? Why?

Graphs of $s(t)$, $p(t)$, and $r(t)$

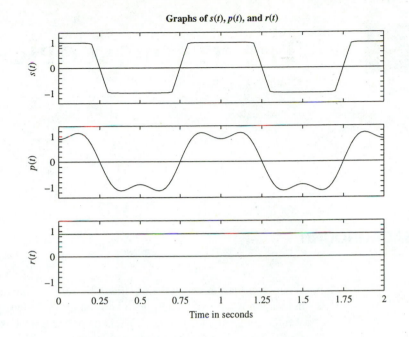

FIGURE 4.4   Graphs of $s(t)$, $p(t)$, and $r(t)$ in Experiment 4

Graphs of $s(t)$ and $p(t)$; $p(t)$ and $r(t)$; $s(t)$ and $r(t)$

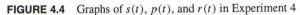

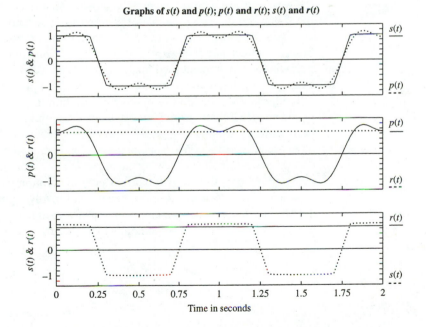

FIGURE 4.5   Comparison graphs in Experiment 4

Experiment 5

THE STROBOSCOPIC EFFECT

BACKGROUND

When you view a wagon being drawn by running horses in a western movie, the wagon wheels often appear to be rotating in the wrong direction and even, at times, not rotating at all. This phenomenon is a stroboscopic effect that is the result of sampling the rotating wagon wheels every T_s seconds. The sampling is a consequence of the film being composed of frames. Each frame is a picture of the wagon at one time instant and the moving picture is obtained by projecting the frames on a screen at a rate of R frames per second. Usually, $R = 1/T_s$. Occasionally, other choices of R are used to obtain special effects. The stroboscopic effect has been exploited to study various objects whose motion is periodic. The instrument used for such studies is called a stroboscope, which is a light source that is pulsed on every T seconds for a short interval. The value of T is adjustable by the user. A moving object illuminated by a stroboscope thus would be seen only for a brief interval every T seconds. In effect, the motion has been sampled at a rate of $F_s = 1/T$ samples per second. Thus, aliasing should be expected when a stroboscope is used. This is one case in which aliasing is desired because it allows viewing the object in slow motion. In this experiment, the stroboscopic effect of two types of periodic motion will be examined. The two examples demonstrated in this experiment involve spatial sampling in two dimensions that is somewhat more complicated than the sampling of time functions. We'll not even consider sampling in three dimensions, which can be rather involved.

EXPERIMENTAL PROCEDURE

▬▬▬▬▬▬▬▬▬ ■ ▬▬▬▬▬▬▬▬▬

The MATLAB function for this experiment is `expr05`. After running `expr05`, you will be asked to choose one of two illustrations: a rotating pendulum or a swinging pendulum.

THE ROTATING PENDULUM

The first illustration is of a pendulum that is rotating similar to a wheel. The rotating pendulum is flashed every $10N$ degrees in which N is a positive integer of your choice. The sampling rate thus is $36K/N$ samples per second in which K is the number of pendulum rotations per second. The rotation rate, K, also can be chosen for your viewing convenience by choosing the value of the parameter I. The relation between K and I is $(mI + b)K = 1$. The values of m and b are computer dependent and can be determined for the computer being used by measuring K for a few values of I.

■ **QUESTION 5.1**

Determine the values of m and b for the computer you are using. For this determination, values of I in the thousands may be required. Discuss your procedure for determining the values of m and b, and compare it with other possible procedures.

A high rotation rate will enable you to observe the stroboscopic effect more easily. To understand the phenomenon, however, a slow rotation rate is required to observe exactly what is happening. We'll examine the stroboscopic effect for $N = N_0$, $N_0 \pm 1$, and $N_0 \pm 2$ for several values of N_0.

Choose $N_0 = 72$ for the first example of a rotating pendulum. Analyze the observed sampled motion in terms of aliasing by expressing the rotating pendulum's horizontal and vertical motions as $x(t) = L\cos(\omega t + \varphi)$ and $y(t) = L\sin(\omega t + \varphi)$. Now repeat with $N_0 = 36$. Analyze and compare the sampled motions for these two cases for the same rotation rate, K. Now analyze and compare the two cases for which $N_0 = 18$ and $N_0 = 54$. To test your understanding, predict what should be observed for other values of N_0 and check your predictions experimentally. For example, what would you predict for $N_0 = 12$ and $N_0 = 9$?

THE SWINGING PENDULUM

The second illustration, that of a swinging pendulum, is somewhat more involved than the rotating pendulum. You will specify the maximum swing angle. Also, the swinging rate can be chosen by the choice of the factor I, as in the rotating-pendulum illustration. The swinging pendulum is flashed $72/N$ times per swing, in which N is a positive integer of your choice. We'll examine the stroboscopic effect for $N = N_0$, $N_0 \pm 1$, and $N_0 \pm 2$.

First choose $N_0 = 72$ and observe the sampled pendulum for a maximum swing angle of 90 degrees. The swinging pendulum for this case can be viewed as the lower half of two rotating pendula that are rotating at the same rate but in opposite directions. Using this model, analyze the sampled motion in terms of aliasing as you did in the rotating-pendulum illustration. Now choose another value for the maximum swing angle. How does the sampled motion compare with the cases studied above? Repeat all the above for the case in which $N_0 =$ 36. To test your understanding of the sampled swinging pendulum, predict what should be observed for the case in which $N_0 = 18$ and check your prediction experimentally.

AN ILLUSTRATION

When expr05 is run, the choice of either a rotating or a swinging pendulum is provided in the MATLAB command window, as shown below.

```
Experiment 5
------------
```

```
THE STROBOSCOPIC EFFECT REALLY IS A FORM OF ALIASING.
TWO CASES ARE ILLUSTRATED IN THIS EXPERIMENT BY FLASHING A PENDULUM.
THE FIRST CASE IS A PENDULUM ROTATING SIMILAR TO A WHEEL.
THE SECOND CASE IS THAT OF A SWINGING PENDULUM.
YOU CAN CHOOSE ONE OF TWO ILLUSTRATIONS:

1. A ROTATING PENDULUM
2. A SWINGING PENDULUM

TYPE WHICH ILLUSTRATION DO YOU DESIRE (1 OR 2)?
```

If, for example, the rotating pendulum is chosen, the following display appears.

```
*************************
*                       *
* THE ROTATING PENDULUM *
*                       *
*************************

THE STROBOSCOPIC EFFECT WILL BE ILLUSTRATED BY "FLASHING" A
ROTATING PENDULUM EVERY 10 * N DEGREES.

THE FLASHING SAMPLE RATE THUS IS 36*K/N FLASHES PER SECOND.
```

```
THE RATE, K, IS INVERSELY PROPORTIONAL TO THE SLOWNESS FACTOR
WHICH IS EQUAL TO A CONSTANT PLUS AN INTEGER, I, OF YOUR CHOOSING.

FOR VIEWING EASE, A FLASHED PORTION WILL REMAIN LIGHTED
ON THE SCREEN UNTIL THE NEXT FLASH.

DESIRED INTEGER VALUE OF N (ENTER 0 TO EXIT) = 34

DESIRED INTEGER VALUE OF I (ENTER 0 TO EXIT) = 50
```

A snapshot of the rotating pendulum for $N = 34$ is shown in Figure 5.1. Note that the number of degrees the pendulum rotates between flashes is given at the top left part of the screen. Since the sequence of flashed positions is periodic, the number of flashes per period and also the number of rotations of the pendulum per period are given at the left edge of the screen.

The Pendulum Simulation

The pendulum is being flashed every 340 degrees.

The flashed sequence is periodic, with 18 flashes per period.

This requires 17 complete rotations of the pendulum.

STOP

FIGURE 5.1   Snapshot of the rotating pendulum for $N = 34$ in Experiment 5

Note that the black circle is the last flash that will remain on the screen until the next flash. The pendulum will continue to rotate and be flashed until you press the stop button. You then can exit the program. If you choose not to exit the program, the first screen will appear.

CHAPTER II

THE DISCRETE-TIME FOURIER TRANSFORM

The experiments in this chapter are concerned with a study of some aspects of the Fourier transform of discrete sequences. Theoretically, the discrete-time Fourier transform (DTFT) of a sequence is[1]

$$X(e^{j2\pi f}) = \sum_{n=-\infty}^{\infty} x(n)\, e^{-j2\pi f n} \tag{II.1}$$

If a computer is used to perform this computation, then only a finite number of terms, N, can be summed. Thus, the computed Fourier transform is

$$X_c(e^{j2\pi f}) = \sum_{n=0}^{N-1} x(n)\, e^{-j2\pi f n} \tag{II.2}$$

Note that $X_c(e^{j2\pi f})$ can be viewed as the DTFT of a truncated version of $x(n)$. Furthermore, the sum can be evaluated for only a finite number of values of the frequency, f. Thus, the computation performed by the computer is

$$X_c(e^{j2\pi f_k}) = \sum_{n=0}^{N-1} x(n)\, e^{-j2\pi f_k n} \tag{II.3}$$

For computational efficiency, the frequencies f_k are chosen to be

$$f_k = \frac{k}{2^P}; \quad k = 0, 1, 2, ..., 2^P - 1 \tag{II.4}$$

in which $2^P \geq N$. This is called the 2^P-point discrete Fourier transform (DFT) of the N-point sequence $x(n)$.[2] Note that the 2^P-point DFT is a sampled version of $X_c(e^{j2\pi f})$ in which the sampling interval is 2^{-P}.

---

[1] The notation used for the Fourier transform of a sequence is chosen to be consistent with that for the z-transform of a sequence. This is the common practice.
[2] In the literature, the DFT, $X_c(e^{j2\pi f_k})$, is usually denoted by $X(k)$.

▪ QUESTION II.1   The DTFT is a transformation of $x(n)$ to $X(e^{j2\pi f})$. Show that the transformation is linear. Is the DFT also a linear transformation?

▪ QUESTION II.2   Why is the DFT of $x(n)$ not evaluated for values of k larger than $2^P - 1$?

An efficient algorithm often used to compute the DFT is called the fast Fourier transform (FFT) algorithm. This particular algorithm is used in MAT-LAB for the computations in this set of experiments in order to minimize the computation time. However, we shall not be concerned with the details of this particular algorithm because we are concerned more with an experimental study of various aspects of Fourier analysis than with efficient computer methods by which the necessary computations can be accomplished.

A graph of the magnitude of $X_c(e^{j2\pi f})$ in eq. (II.2), $\left|X_c(e^{j2\pi f})\right|$, will be displayed in the figure window. The graph is obtained by connecting the computed values of the magnitude of the DFT, $\left|X_c(e^{j2\pi f_k})\right|$, by straight lines. (Note that this is a first-order polynomial D/A conversion of the DFT in accordance with our discussion in Experiment 3.) To obtain good screen graphs, P in eq. (II.4) has been chosen equal to 10 so that $X_c(e^{j2\pi f})$ is evaluated at 512 values of f between 0 and $\frac{1}{2}$.

Experiment 6

RESOLUTION OF TWO SINUSOIDS

BACKGROUND

The Fourier transform of a time function, $s(t)$, is

$$S(j2\pi F) = \int_{-\infty}^{\infty} s(t)e^{-j2\pi F}\, dt \qquad (6.1)$$

A computer cannot be used to evaluate this expression exactly because a computer can process only discrete sequences, $x(n)$. However, an approximation of $S(j2\pi F)$ can be obtained by determining the DTFT, $X(e^{j2\pi f})$, of a sequence, $x(n)$, obtained by sampling $s(t)$. As we discussed in the introduction, the DTFT cannot be determined by a computer. Only the DFT, $X_c(e^{j2\pi f_k})$, which is an approximation of the DTFT, can be determined. The larger the values of N and P in eqs. (II.3) and (II.4), the smaller is the approximation error. The value of P determines only the interpolation error in the first-order polynomial D/A conversion. The value of P used in this chapter was chosen so that the interpolation error is of the same order as the screen quantization error. Consequently, the displayed plot can be considered to be $\left|X_c(e^{j2\pi f})\right|$. From our discussion in the introduction, observe that $X_c(e^{j2\pi f})$ can be viewed as the DTFT of a truncation of $x(n) = s(nT_s)$. For this experiment, we shall study the effect of N on the approximation error for the case in which

$$s(t) = \cos(2\pi F_1 t + \theta) + A\cos(2\pi F_2 t) \qquad (6.2)$$

EXPERIMENTAL PROCEDURE

■

PART 1

We begin by first examining $X_c(e^{j2\pi f})$ given by eq. (II.2) for the case in which $s(t)$ is a single sinusoid; this is obtained by choosing $A = 0$ in eq. (6.2). Before executing the MATLAB function for this case, derive an expression for $X_c(e^{j2\pi f})$ for the case in which $A = 0$. Note that $X_c(e^{j2\pi f})$ can be expressed as $X_a(e^{j2\pi f}) + X_b(e^{j2\pi f})$ in which $\left|X_a(e^{j2\pi f})\right|$ has a peak at $f = f_1$ and $\left|X_b(e^{j2\pi f})\right|$ has a peak at $f = -f_1$. Express $X_a(e^{j2\pi f})$ and $X_b(e^{j2\pi f})$ in polar form. Sketch a graph of $\left|X_a(e^{j2\pi f})\right|$ and $\left|X_b(e^{j2\pi f})\right|$ in the range $0 \le f \le 0.5$.

■ **QUESTION 6.1**   At what values of f is each graph zero?

■ **QUESTION 6.2**   Describe the specific effect of the parameters N, F_1, F_s, and θ on each graph.

■ **QUESTION 6.3**   For what values of $f \ge 0$ is $\left|X_c(e^{j2\pi f})\right| \approx \left|X_a(e^{j2\pi f})\right|$? Note that this frequency range is a function of the parameters N, F_1, F_s, and θ.

Begin this experiment by running `expr06` from the MATLAB command window. You will be asked to specify the parameters given in eq. (6.2):

- Sampling rate, F_s, in samples per second
- Frequency, F_1, in hertz
- Phase, θ, in degrees
- Amplitude, A
- Difference frequency, $dF \triangleq F_2 - F_1$, in hertz
- Number of samples of $x(n)$, N

Now verify your theoretical result by comparing it with the MATLAB graph. For this comparison, choose $A = 0$, $F_s = 200$ samples per second, and $N = 25$ samples. Obtain the screen graph of $\left|X_c(e^{j2\pi f})\right|$ versus f for $F_1 = 0$, 17, 50, 80, 150, and 183 Hz and various values of θ.

■ **QUESTION 6.4**   Are any of the graphs identical? If so, explain why. Does this agree with your theoretical result?

■ **QUESTION 6.5**   What is the specific effect of θ on the graphs? Is this in agreement with your theoretical result? Specifically discuss the effect at $f = 0$ and $f = F_1/F_s$.

PART 2

A problem that arises in many fields of scientific work, such as spectroscopy, radio astronomy, ocean wave analysis, and geophysics, is the resolution of sinusoids of which a waveform is composed. In this part, we shall study some basic aspects of this problem. Use $A = 1$ throughout Part 2 so that $s(t)$ is composed of two equal-amplitude sinusoids. First, use the theoretical expression obtained in Part 1 and the fact that the DTFT is a linear transformation of functions of n to functions of f to obtain an expression for $X_c(e^{j2\pi f})$ for the case in which $A = 1$. Express $X_c(e^{j2\pi f})$ as $X_a(e^{j2\pi f}) + X_b(e^{j2\pi f})$ in which $X_a(e^{j2\pi f})$ has peaks at f_1 and f_2 and $X_b(e^{j2\pi f})$ has peaks at $-f_1$ and $-f_2$.

■ **QUESTION 6.6**   In a plot of $\left|X_c(e^{j2\pi f})\right|$, determine the values of N, F_1, F_2, F_s, and θ for which $X_b(e^{j2\pi f})$ can be ignored for f in the range between and including the peaks at f_1 and f_2.

■ **QUESTION 6.7**   Use your expression for $X_a(e^{j2\pi f})$ to determine the value of $\theta = \theta_0$ in degrees for which $\left|X_a(e^{j2\pi f})\right|$ is a maximum at $f = f_0 = (f_1 + f_2)/2$.

Express $X_a(e^{j2\pi f})$ as $X_1(e^{j2\pi f}) + X_2(e^{j2\pi f})$ in which $X_1(e^{j2\pi f})$ and $X_2(e^{j2\pi f})$ are in polar form and $\left|X_1(e^{j2\pi f})\right|$ has a peak at $f = f_1$ and $\left|X_2(e^{j2\pi f})\right|$ has a peak at $f = f_2$.

Now choose $F_s = 200$ samples/second, $F_1 = 48$ Hz, and $N = 50$. With these values, obtain the screen graphs for each of the following sets of parameters of θ and $dF = F_2 - F_1$:

1. $dF = 4$ with $\theta = 0, 90, 180$ degrees, and θ_0 determined above
2. $dF = 8$ with $\theta = 0, 90, 180$ degrees, and θ_0 determined above

Use your expression for $X_c(e^{j2\pi f})$ derived above to explain the shape of $\left|X_c(e^{j2\pi f})\right|$ observed on the screen. Specifically discuss the frequency region $f_1 \leq f \leq f_2$.

Note that if dF is sufficiently large, then, for any value of θ, there are two maxima of $\left|X_c(e^{j2\pi f})\right|$: one at $f = f_1$ corresponding to F_1 and one at $f = f_2$ corresponding to F_2. However, there is a dF_{min} such that these two maxima cannot be distinguished for some values of θ if $dF \leq dF_{min}$. Generally, in resolving sinusoids of a waveform, we do not know the relative phase, θ. Thus, we desire to determine dF_{min} for which we are assured that the sinusoids can just be resolved for any value of θ. Determine the value of dF_{min} experimentally for the case in which $F_s = 200$ samples per second, $F_1 = 50$ Hz, and $N = 50$. Note that the worst case is for $\theta = \theta_0$.

This one result can be used to obtain a general expression for dF_{min} that is valid for any set of values of the parameters F_1, F_2, F_s, and N. The desired expression is a function of the parameters that can be obtained from an examination of your expression for $X_c(e^{j2\pi f})$ by noting that the graph of $\left|X_c(e^{j2\pi f})\right|$ is the same if the value of a certain function of the parameters is the same and evaluating the function using the values of your experiment. Use your expression for $X_c(e^{j2\pi f})$ together with your experimental result to obtain a general expression for dF_{min} in terms of the parameters F_1, F_2, F_s, and N for the case in which $A = 1$. Verify your expression experimentally and discuss your choice of sets of parameter values used for this verification.

Note that you have been able to obtain a general expression from the experimental result obtained for one set of parameter values. This technique is used in many scientific problems that are difficult to analyze theoretically. The procedure is to form dimensionless functions of the parameters involved. Experimental results then are obtained in terms of these dimensionless functions. You should check the dimensions of your function of the parameters for dF_{min}. This approach is a very powerful one and is one of the reasons dimensional analysis is an important scientific topic.

Please note that you have just determined conditions for resolving two equal-amplitude sinusoids. Resolving the two sinusoids does not necessarily imply that the two sinusoids can be separated with small error. The conditions for their separation will be examined in the next experiment.

AN ILLUSTRATION

The MATLAB script that appears when you run `expr06` is shown below for parameter values $F_s = 40$ samples per second, $F_1 = 8$ Hz, $\theta = 0$ degrees, $A = 1$, $dF = 4$ Hz (so that $F_2 = F_1 + dF = 12$ Hz), and $N = 40$ samples. Remember that all parameter values must be entered in decimal form.

EXPERIMENT 6

```
    THE CONTINUOUS WAVEFORM TO BE SAMPLED IS
  s(t) = cos(2*pi*F1*t + Theta) + A*cos(2*pi*(F1+dF)*t
THE SEQUENCE x(n) CONTAINS N SAMPLES OBTAINED BY SAMPLING
s(t) AT A RATE OF Fs SAMPLES PER SECOND FROM t=0 THROUGH
t=(n-1)/fS.  THE DTFT OF x(n) WILL BE DETERMINED AND A
GRAPH OF THE DTFT MAGNITUDE WILL BE DISPLAYED.


    Sampling Rate, Fs, in sam/sec = 40
          Frequency, F1, in hertz = 8
        Phase, Theta, in degrees = 0
                  Amplitude, A = 1
Difference frequency, dF, in hertz = 4
    Number of samples (<= 100), N = 40
```

After all the required data are entered, the graph of $\left|X_c(e^{j2\pi f})\right|$ will appear in a MATLAB figure window. An example of a graph obtained with the given values is shown in Figure 6.1. The chosen parameter values are displayed at the top of the graph. The graph is ten divisions high. The number of units per division on the ordinate is determined by the value of C, which is given at the top of the graph. For the example, $C = 20$, the ordinate scale is 2 units per division for the graph shown. Note that the two peaks on the graph are well separated at $f = 0.2$ and 0.3 cycles/sample. For other values of the parameters, the region between the peaks is different so that they may not be discernible.

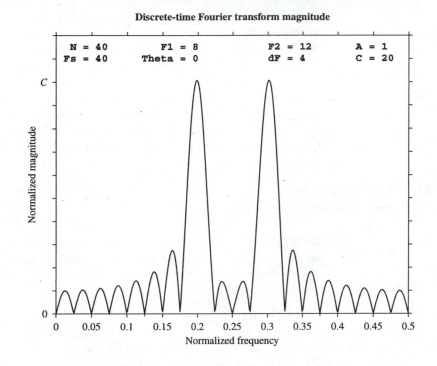

FIGURE 6.1   Plot of the DTFT magnitude in Experiment 6

SEPARATION OF TWO SINUSOIDS

BACKGROUND

Some basic aspects of resolving two sinusoids were investigated in the last experiment. However, even though two sinusoids can be resolved, it may not be possible to separate them with a small error. The considerations involved in the separation of sinusoids are basic to many problems in communications. In this experiment, one technique using the computer for separating them will be investigated.

For this study, we shall try to eliminate the second sinusoidal term, $A \cos(2\pi F_2 t)$, from the signal

$$s(t) = \cos(2\pi F_1 t + \theta) + A \cos(2\pi F_2 t) \tag{7.1}$$

If the computed spectra of the two sinusoids did not overlap, then, as illustrated in Figure 7.1, we could eliminate the second sinusoid by multiplying $X_c(e^{j2\pi f})$ by $H(e^{j2\pi f})$ in which $H(e^{j2\pi f})$ equals zero over a frequency range corresponding to the DTFT spectrum of $\cos(2\pi F_2 t)$ and equals one for other frequencies. The filtered waveform is then the inverse discrete-time Fourier transform (IDTFT) of the product, $X_c(e^{j2\pi f})H(e^{j2\pi f})$. This is the system implemented in the MATLAB function expr07. However, the DTFT spectra of the two sinusoids do overlap, so the system shown in Figure 7.1 is an approximation of an ideal analog band-reject filter implemented using discrete systems.

If the spectra do overlap, perfect sinusoidal separation using this system is not possible and there will be an error, $e(t)$, given by

$$e(t) = r(t) - \cos(2\pi F_1 t + \theta) \tag{7.2}$$

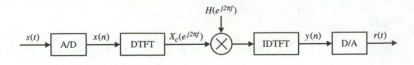

FIGURE 7.1 Ideal analog band-reject filter using discrete systems

EXPERIMENTAL PROCEDURE

To begin this experiment, run `expr07` from the MATLAB command window. The parameters that must be specified are the following:

- The sampling rate, F_s
- The number of samples taken, N
- The sinusoidal frequencies, F_1 and F_2
- The phase, θ
- The amplitude, A

A screen graph of $\left|X_c(e^{j2\pi f})\right|$ will be displayed after the parameter values are entered. From a study of this graph, the frequency range for which $H(e^{j2\pi f}) = 0$ is then chosen. The product $\left|H(e^{j2\pi f})X_c(e^{j2\pi f})\right|$ will be displayed. An opportunity to change the frequency range is available if this product is not acceptable. Upon acceptance, four sequence plots will be displayed:

1. $x(n) = s(nT_s)$
2. $x_1(n) = \cos(2\pi f_1 nT_s)$
3. $e(n) = y(n) - \cos(2\pi f_1 nT_s)$
4. $y(n)$

In addition, the maximum amplitude of each sequence and the root-mean-square (rms) value of the error sequence, $e(n)$, will be given.

PART 1

First, choose the parameter values to be $F_s = 200$ samples per second, $N = 100$ samples, $F_1 = 10$ Hz, $F_2 = 90$ Hz, $\theta = 0$ degrees, and $A = 1$. Determine the optimum reject band, $F_L < F < F_H$, for which the error is the smallest. For this, a criterion for the size of the error must be chosen. The maximum amplitude and the rms value of $e(n)$ are given for your use. However, these numbers cannot be used without some consideration. Observe $e(n)$ for some selections of the reject band. Note that $e(n)$ for a particular case can be rather small for most values of n, but its rms value can be rather large due to a few large values of $e(n)$ near the ends of the sequence. Thus, to determine the optimum reject band, careful consideration first must be given to the choice of a meaningful criterion for the size of the error.

The chosen criterion must be quantitative. That is, the criterion must result in a number being assigned to the error size so that the various error sequences can be ordered according to size. In mathematics, this is called a norm of $e(n)$ and is denoted by $\|e(n)\|$. If your norm of $e(n)$ is to be a proper one, it must have the following properties:

1. $\|e(n)\| \geq 0$
2. $\|e(n)\| = 0$ if and only if $e(n) = 0$
3. $\|\alpha e(n)\| = |\alpha| \|e(n)\|$ in which α is any constant
4. $\|e_1(n) + e_2(n)\| \leq \|e_1(n)\| + \|e_2(n)\|$

■ QUESTION 7.1   Discuss the physical significance of the four properties of a norm.

■ QUESTION 7.2   Show that the rms value of $e(n)$ and also the maximum value of $|e(n)|$ are proper norms of $e(n)$. Is α times the rms value of $e(n)$ plus β times the maximum value of $|e(n)|$, in which α and β are constants, a proper norm of $e(n)$?

Note that different choices for $\|e(n)\|$ can result in different orderings of the error sequences. That is, the "optimum" result or solution of a problem is, to a large extent, a consequence of the choice of the norm.

■ QUESTION 7.3   Choose a norm for $e(n)$ and discuss why you believe your choice is a meaningful one for this experiment.

Use your definition of $\|e(n)\|$ to experimentally determine the optimum reject band for the set of parameters specified above and discuss its correspondence with your intuitive expectation.

PART 2

In Part 1, the DTFT spectra of the two sinusoids do not overlap very much and your result should be close to your intuitive expectation. To consider a case in which the DTFT spectra of the two sinusoids do overlap, choose the parameter values to be $F_s = 100$ samples per second, $N = 50$ samples, $F_1 = 5$ Hz, $F_2 = 9$ Hz, $\theta = 0$, and $A = 1$. Now choose $F_L = 7$ Hz and plot a graph of the maximum value of $|e(n)|$, the rms value of $e(n)$, and your $\|e(n)\|$ versus F_H. Discuss the graphs of the three norms you obtained. Compare the optimum values of F_H for the three norms. Plot each of the three norms of $e(n)$ versus the phase angle, θ, for $F_L = 7$ Hz and F_H equal to the optimum value determined above. Discuss the curves you obtained.

PART 3   In many applications, such as spectroscopy, it is only necessary to resolve the sinusoids of which a waveform is composed. However, in communications, it is often necessary to actually separate them. We mentioned in Experiment 6 that even though two sinusoids can be resolved, it may not be possible to separate them with small error. Investigate this by choosing a case in which the two sinusoids can just be resolved and study the separation error. By what factor must the frequency difference, dF_{min}, be increased in order that $\|e(n)\|$ be acceptably small? Discuss your choice of an acceptably small value of $\|e(n)\|$.

AN ILLUSTRATION

The MATLAB script that is obtained when you run expr07 in the command window is the informational one shown next.

                    EXPERIMENT 7

THE SEQUENCE x(n) IS OBTAINED BY SAMPLING THE CONTINUOUS WAVEFORM

    s(t) = cos(2*pi*F1*t + Theta) + A*cos(2*pi*F2*t)

AT A RATE OF Fs SAMPLES PER SECOND FROM t = 0 THROUGH t = (N-1)/Fs.
A GRAPH OF THE DTFT MAGNITUDE WILL BE DISPLAYED. PRESS "ENTER" TO
CHOOSE THE REJECT BAND, [FL, FH], OF THE BAND REJECT FILTER.   THEN
PRESS "ENTER" TO VIEW A GRAPH OF THE DTFT MAGNITUDE OF THE FILTERED
SEQUENCE y(n). UPON ACCEPTANCE OF YOUR CHOICE OF THE REJECT BAND,
A DISPLAY OF FOUR GRAPHS WILL BE SHOWN:

  1. THE ORIGINAL SEQUENCE, x(n)   2. THE DESIRED SEQUENCE, x1(n)

  3. THE ERROR SEQUENCE,   e(n)   4. THE FILTERED SEQUENCE, y(n)

*** PRESS ENTER TO CONTINUE ***

When you press Enter, the next script that appears in the command window is the one shown below for entering all the desired signal parameters. The screen is shown for parameter values $F_s = 40$ samples per second, $N = 40$ samples, $F_1 = 8$ Hz, $F_2 = 12$ Hz, $\theta = 0$, and $A = 1$.

THE CONTINUOUS WAVEFORM TO BE SAMPLED IS

    s(t) = cos(2*pi*F1*t + Theta) + A*cos(2*pi*F2*t)

```
Sampling rate, Fs, in sam/sec = 40
Number of samples (<= 200), N = 40
       Frequency, F1, in hertz = 8
       Frequency, F2, in hertz = 12
     Phase, Theta, in degrees = 0
                  Amplitude, A = 1
```

A graph of the magnitude of the DTFT, $\left|X_c(e^{j2\pi f})\right|$, as shown in Figure 7.2, appears in the MATLAB figure window after you enter the parameter values and press Enter. Note that the value of the graph maximum, C, is given at the top of the graph. For the illustration, $C = 20$. Since there are ten divisions between 0 and C, there are $20/10 = 2$ units per division for the graph shown. This graph is the same as that illustrated in Experiment 6.

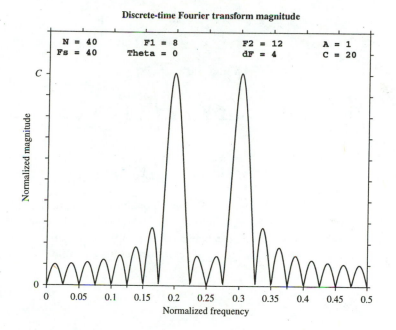

FIGURE 7.2   Plot of the DTFT magnitude in Experiment 7

This figure window will remain on the screen until the left mouse button is pressed *on the figure window*. Then the screen shown below appears. The values of F_L and F_H must now be specified.

```
CHOOSE THE REJECT BAND, [FL, FH], OF THE IDEAL FILTER.
     BE CERTAIN THAT 0 <= FL <= FH <= Fs/2.


  Desired value of FL in hertz = 9
  Desired value of FH in hertz = 20
```

If, for example, $F_L = 9$ Hz and $F_H = 20$ Hz were chosen as shown, then a graph of the filtered spectrum shown in Figure 7.3 appears.

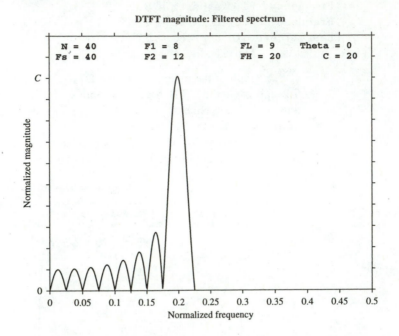

DTFT magnitude: Filtered spectrum

FIGURE 7.3   Plot of the filtered spectrum in Experiment 7

This graph will remain on the screen until the left mouse button is pressed on the figure window. The IDTFT will then be computed and the four graphs of $x(n)$, $x_1(n)$, $e(n)$, and $y(n)$ will appear in a MATLAB figure window. They are shown in Figure 7.4 for the preceding example.

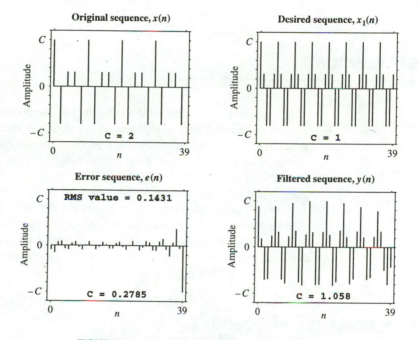

FIGURE 7.4   The time-domain plots in Experiment 7

Experiment 8

ANALYSIS OF PERIODIC WAVEFORMS

BACKGROUND

The resolution of two sinusoids was studied in Experiment 6. However, as discussed in Experiment 4, no interesting waveform is bandlimited, so sampling it always will result in aliasing. Thus, aliasing is a major concern in the determination of a spectrum with a digital computer. Of course, as in Experiment 4, this effect could be reduced by prefiltering, but prefiltering entails extra complexity and other errors. In this experiment, we shall analyze the spectra of two nonsinusoidal periodic waveforms, a square wave and a triangular wave, without prefiltering. First, F_s samples per second of the periodic waveform, $s(t)$, are taken to obtain $x(n) = s(nT_s - D)$ in which D is a sampling delay. The DFT using N samples of the sequence $x(n)$ is then determined and a graph of $\left| X_c(e^{j2\pi f_k}) \right|$ is displayed on the screen.

EXPERIMENTAL PROCEDURE

This experiment is started by running the function `expr08` from the MATLAB command window. There are three waveforms used for investigation in this experiment; these investigations are described below.

PART 1

To begin, study the effect of N, the number of samples used in determining $X_c(e^{j2\pi f})$. For this, choose waveform A for which $s(t)$ is a sinusoid with frequency $F = F_s/4$ Hz and a phase of θ degrees. The frequency was chosen so that the computed spectrum is centered on the graph.

■ **QUESTION 8.1**   How many samples per period of the sinusoid are taken? What is the effect of F_s on the computed graph?

Is the spectrum magnitude affected by θ? For this study, choose $F_s = 20$ samples per second and $N = 5$ samples. Examine $\left|X_c(e^{j2\pi f})\right|$ for various values of θ. From this examination, it is clear that θ can have a significant effect on the computed spectrum. Discuss the effect you have observed. Now, with $F_s = 20$, examine $\left|X_c(e^{j2\pi f})\right|$ as a function of N. Discuss the effect of θ on the spectrum as a function of N. What is the effect of N on the shape of the spectrum? Discuss the effect of N you have observed and theoretically explain your observations.

PART 2   Now that the effect of N on the calculated spectrum of a sinusoid is known, we will examine the calculated spectrum of a square wave, $s(t)$, with a fundamental period of T seconds as shown in Figure 8.1. The DFT is determined using N samples of $x(n) = s(nT_s - D)$. The effect of the parameters F_s, N, T, and D on the calculated spectrum will be studied.

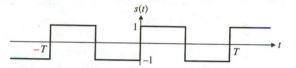

FIGURE 8.1   Square waveform used in Experiment 8

First, choose $N = 200$ samples, $F_s = 200$ samples per second, $T = 0.1$ second, and $D = 0$. How does $\left|X_c(e^{j2\pi f})\right|$ compare with your theoretical expectation? For example, compare the position and amplitude of each harmonic peak. Does aliasing account for any observed differences? Also, compare the width of each harmonic peak with the results of your study of waveform A. Now vary one parameter at a time and study its effect on the calculated spectrum. Does D affect the spectrum? Now vary N from about 180 to 220. Does a 10% change of N affect the spectrum? Now vary the sampling rate, F_s. For example, is the spectrum changed if $F_s = 190$ samples per second? Discuss and explain any observed changes.

Often, the exact period of a periodic waveform is not known with precision. Note from your results above that aliasing can be a problem if the sampling parameters are not chosen properly. However, if the amplitudes of the aliased harmonics are small, then they may not cause significant error in the regions of the spectrum where they occur. For example, choose $N = 200$ samples, $F_s = 400$ samples per second, $T = 0.1$ second, and $D = 0$. Then examine the spectrum with $F_s = 390$. Discuss and explain any observed changes.

From your studies, what values of the sampling parameters should be used to determine the first three nonzero harmonics of the square wave without significant error. Assume that the square wave fundamental period is known only within 10% of its correct value. Check your conclusion experimentally.

PART 3

The square wave was examined first because, in a sense, it is a worst case. As a function of the harmonic number, the harmonic amplitudes of the triangular waveform decrease at a much faster rate than those of the square wave. The triangular waveform, $s(t)$, used is shown in Figure 8.2. The fundamental period of $s(t)$ is T.

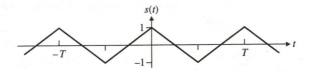

FIGURE 8.2   Triangular waveform used in Experiment 8

■   **QUESTION 8.2**   Compare the Fourier series of the triangular wave shown with that of the square wave shown in Figure 8.1.

The DTFT is calculated using N samples of $x(n) = s(nT_s - D)$. Theoretically, what values of the sampling parameters should be used to determine the first three nonzero harmonics of the triangular wave without significant error? Assume that the triangular wave fundamental period is known only within 10% of its correct value. Experimentally check your conclusion and compare the required values with those required for the square wave.

From the results of this experiment, note that it is possible to study the spectrum of a periodic waveform experimentally without prefiltering. If N is sufficiently large so that the main lobes are narrow, then, with a proper choice of F_s, the harmonic frequencies that are aliased do not overlap with those that are not aliased. In this manner, one can measure the amplitudes of the harmonic frequencies that are not aliased and also some that are aliased.

AN ILLUSTRATION

The MATLAB script that is displayed when you execute `expr08` in the command window is the informational one shown below.

```
                    EXPERIMENT 8

THE DTFT OF A SEQUENCE x(n) WILL BE DETERMINED AND A GRAPH
```

OF THE DTFT MAGNITUDE WILL BE DISPLAYED. THE SEQUENCE x(n) CONTAINS
N SAMPLES OBTAINED BY SAMPLING ONE OF THE CONTINUOUS PERIODIC
WAVEFORMS DESCRIBED BELOW AT A RATE OF Fs SAMPLES/SEC.

    A. s(t) = cos(0.5*pi*Fs*t + Theta)

    B. s(t) = A SQUARE WAVE WITH AN AMPLITUDE ONE
              AND A FUNDAMENTAL PERIOD OF T SECONDS.

    C. s(t) = A TRIANGULAR WAVE WITH AN AMPLITUDE ONE
              AND A FUNDAMENTAL PERIOD OF T SECONDS.

*** PRESS ENTER TO CONTINUE ***

When you press Enter, a menu screen appears from which you can choose one of the three waveforms. If waveform A is chosen, then the MATLAB script to choose its parameters is displayed in the command window. An example of this script with the parameters $F_s = 40$ samples per second, $N = 40$ samples, and $\theta = 0$ degrees is shown next.

THE CONTINUOUS WAVEFORM TO BE SAMPLED IS

    s(t) = cos(0.5*pi*Fs*t + Theta)

  Number of samples (0 < N <= 1000), N = 40
  Sampling rate, Fs, in samples/second = 40
            Phase, Theta, in degrees = 0

After you enter the parameters, the DTFT graph shown in Figure 8.3 appears. Note that the vertical scale is C units per ten divisions. As given at the top of the graph, C is 20, so the scale is 2 units per division for the example illustrated.

If waveform B or C is chosen, then the previous command window script appears without the parameter Phase, Theta, in degrees but with the addition of the required parameters: THE FUNDAMENTAL PERIOD OF S(t) IN SECONDS, T, and THE SAMPLING DELAY IN SECONDS, D. If, for example, waveform B, the square wave, is chosen with the parameters $F_s = 40$ samples per second, $N = 40$ samples, $T = 0.4$ second, and $D = 0$ seconds, then the graph shown in Figure 8.4 appears. Note that all the parameters are listed at the top of the graph.

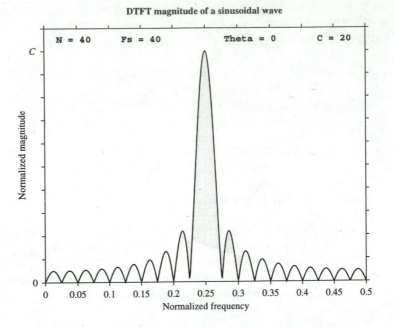

FIGURE 8.3   The DTFT of a sinusoidal waveform in Experiment 8

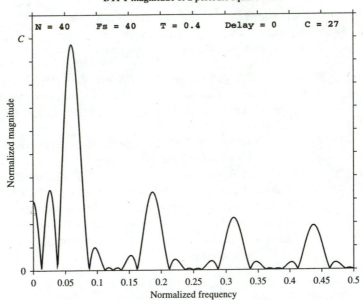

FIGURE 8.4   The DTFT of a square waveform in Experiment 8

CHAPTER III

GAIN AND PHASE-SHIFT

A topic of central importance in the theory of stable, linear, shift-invariant (LSI) discrete systems is the system gain and phase-shift, which we shall examine experimentally. The importance of this topic derives from the fact that the phasor is a characteristic function of a stable LSI discrete system. If the input is

$$x(n) = e^{j\omega n} \tag{III.1}$$

then the system response of a bounded input–bounded output (BIBO) stable system is

$$y(n) = H(e^{j\omega})e^{j\omega n} \tag{III.2}$$

in which $H(e^{j\omega})$, the system transfer function (also called the frequency response function) is given by

$$H(e^{j\omega}) = \sum_{n=-\infty}^{\infty} h(n)e^{-j\omega n} \tag{III.3}$$

The phasor $e^{j\omega n}$ is a characteristic function of a stable LSI system because, when the input is a phasor, the output is equal to the input times a constant, with the constant being the system transfer function.

■ QUESTION III.1   Use convolution to derive $y(n)$ given by eq. (III.2).

Since the system is linear, it immediately follows that if

$$x(n) = \sum_{k} c_k e^{j\omega_k n} \tag{III.4}$$

then

$$y(n) = \sum_{k} c_k H(e^{j\omega_k})e^{j\omega_k n} \tag{III.5}$$

We thus note that if an input can be expressed as a linear combination of phasors, then the output can be determined immediately with the use of the transfer function. This is the reason Fourier analysis is important in the study of LSI discrete systems.

A sinusoid can be expressed as the sum of phasors in the form

$$x(n) = A\cos(\omega n + \varphi)$$
$$= \frac{A}{2}e^{j\varphi}e^{j\omega n} + \frac{A}{2}e^{-j\varphi}e^{-j\omega n} \qquad \text{(III.6)}$$

The corresponding system response is, in accordance with eq. (III.5),

$$y(n) = \frac{A}{2}e^{j\varphi}H(e^{j\omega})e^{j\omega n} + \frac{A}{2}e^{-j\varphi}H(e^{-j\omega})e^{-j\omega n} \qquad \text{(III.7)}$$

Now, $H(e^{-j\omega}) = H^*(e^{j\omega})$ since $h(n)$ is a real function of n. Consequently, the two terms summed in eq. (III.7) are conjugates so that

$$y(n) = 2\operatorname{Re}\left\{ \frac{A}{2}e^{j\varphi}H(e^{j\omega})e^{j\omega n} \right\} \qquad \text{(III.8)}$$

By expressing the transfer function in polar form, we can express this as

$$y(n) = A\left|H(e^{j\omega})\right|\cos\left[\omega n + \varphi + \theta(\omega)\right] \qquad \text{(III.9)}$$

in which $\left|H(e^{j\omega})\right|$ is the system gain and $\theta(\omega) \triangleq \angle H(e^{j\omega})$ is the system phase-shift. [1]

It then follows from the system linearity that if the input can be expressed as a linear combination of sinusoids,

$$x(n) = \sum_k A_k \cos\left[\omega_k n + \varphi_k\right] \qquad \text{(III.10)}$$

then, with the use of the transfer function, the system response can immediately be determined as

$$y(n) = \sum_k A_k \left|H(e^{j\omega_k})\right|\cos\left[\omega_k n + \varphi_k + \theta(\omega_k)\right] \qquad \text{(III.11)}$$

■ QUESTION III.2   Show that $H(e^{-j\omega}) = H^*(e^{j\omega})$ using both eq. (III.3) and the fact that $h(n)$ is a real function of n.

■ QUESTION III.3   Supply the details for obtaining eq. (III.9) from eq. (III.8).

[1] In some literature, $\left|H(e^{j\omega})\right|$ is called the magnitude response and $\theta(\omega)$ is called the phase response.

The z-transform of the unit sample response, $h(n)$, is defined as the power series

$$H(z) = \sum_{n=-\infty}^{\infty} h(n)z^{-n} \qquad \text{(III.12)}$$

If the system is BIBO stable, then the region of absolute convergence for $h(n)$ includes the unit circle in the complex z-plane. For BIBO stable systems, we thus can let $z = e^{j\omega}$ in eq. (III.12). This substitution results in the transfer function, as seen by comparing with eq. (III.3).

There is a z-plane geometric view by which the system gain and phase-shift can be determined. Let p be a point on the unit circle in the complex z-plane that is at an angle of ω radians with the positive real z-axis. Then, if $H(z)$ is a rational function, the system gain at the frequency ω is equal to a constant times the product of the distances from the zeros to the point p divided by the product of the distances from the poles to the point p. Also, the phase-shift is equal to a constant plus the sum of the angles from the zeros to the point p minus the sum of the angles of the poles to the point p. For example, if

$$H(z) = A\frac{z-3}{z+0.5}$$

then, as shown in the figure below, the system gain at the frequency $\omega = \pi/2$ is

$$\left|H(e^{j\pi/2})\right| = |A|\,\frac{\sqrt{10}}{\sqrt{5}/2} = 2\sqrt{2}\,|A|$$

and the phase-shift at the frequency $\omega = \pi/2$ is

$$\theta(\pi/2) = \angle A + [\pi - \arctan(1/3)] - \arctan(2)$$

$$= \angle A + \pi/2 + \arctan(3) - \arctan(2)$$

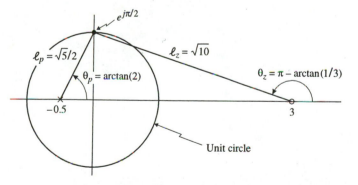

QUESTION III.4   Show that $p = e^{j\omega}$ is a point in the z-plane that is a unit distance from the origin and at an angle of ω radians from the positive real axis.

■ QUESTION III.5   Let z_0 be a point in the z-plane. Show that $\left|e^{j\omega} - z_0\right|$ is equal to the distance from the point z_0 to the point p.

■ QUESTION III.6   Show that the angle of $(e^{j\omega} - z_0)$ is equal to the angle of the line from z_0 to p.

■ QUESTION III.7   Use Questions III.4, III.5, and III.6 to prove the z-plane geometric view of the gain and phase-shift given in the last paragraph above.

The geometric view of the gain and phase-shift is useful in determining without much calculation the approximate filter characteristics from knowledge of the system pole and zero locations. In filter design, the problem is to determine the required pole and zero locations to achieve a desired gain and phase-shift. The understanding gained from the z-plane geometric view, concerning how the locations of the poles and zeros affect the system gain and phase-shift, thus lends a great deal of insight required for filter design. It is this view that we will develop.

Experiment 9

THE MOVING AVERAGE FILTER

BACKGROUND

The objective of this experiment is to study the sinusoidal response of a simple discrete system using the concepts discussed in the introduction to this chapter. The system simulated by the MATLAB function for this experiment is shown in Figure 9.1. The input signal is the sinusoid

$$s(t) = A\cos(2\pi Ft + \gamma) \tag{9.1}$$

which is sampled at the time instances $t_n = nT_s$. The response of the A/D converter thus is

$$x(n) = A\cos(2\pi fn + \varphi) \tag{9.2}$$

The discrete system is an L-point moving average filter whose output, $y(n)$, is the average of the last L values of its input, $x(n)$, so that

$$y(n) = \frac{1}{L}\sum_{k=0}^{L-1} x(n-k) \tag{9.3}$$

Since the filter is a linear, shift-invariant (LSI) discrete system, its response, $y(n)$, to the given input is

$$y(n) = B\cos(2\pi fn + \psi) \tag{9.4}$$

EXPERIMENTAL PROCEDURE

Do the following steps:

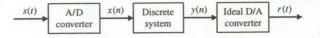

FIGURE 9.1   The system used in Experiment 9

1. Draw a direct form realization of the filter described by eq. (9.3).

2. Determine the system function, $H(z)$, of the filter and sketch the z-plane pole-zero diagram for $L = 4$ and for $L = 9$.

3. Determine and plot the gain, B/A, and the phase-shift, $\theta = \varphi - \psi$, for $L = 4$ and for $L = 9$.

4. Verify your theoretical results experimentally at several frequencies. To do this, run `expr09` in the MATLAB command window. You will be asked to specify the values of A, F, γ, T_s, L, and N. A screen graph of $s(t)$ and $r(t)$ will then be plotted with the abscissa marked at the time instances $t_n = nT_s$. A total of $2N$ samples ($N > L$) will be taken. Only the interval $N T_s \le t \le 2N T_s$ of $r(t)$ will be displayed so as not to include the transient part of the filter response.

   List your experimental results, theoretical results, and the percentage error in tabular form. Note that the phase difference between two sinusoids is best determined by comparing their zero crossings. Make your measurements from the screen graphs about the region $n = 3N/2$ in order to minimize the errors observed in Experiment 2. State why you chose the particular values of A, F, γ, T_s, and N you used for verification. Discuss your results.

5. Obtain a printout for the case in which $A = 1$, $F = 1.0$ Hz, $\gamma = 0$, $T_s = 0.125$ second, $L = 4$, and $N = 25$. Show on the plot how you determined B/A and θ for this case. Do not forget to show how you determined the sign of θ.

6. Obtain a printout for the case in which $F = 7$ Hz and the other parameter values are the same as those specified in step 5.

7. Compare the experimental results you obtained in steps 5 and 6. Use sampling theory to explain any similarities and differences.

AN ILLUSTRATION

The MATLAB script that appears when you run `expr09` is shown below for parameter values $A = 1$, $F = 2$ Hz, $\gamma = 0$ degrees, $T_s = 0.1$ second, $N = 21$ samples, and order $L = 3$.

EXPERIMENT 9

2N SAMPLES OF THE CONTINUOUS SIGNAL

    s(t) = A*cos(2*pi*F*t+Gamma)

ARE TAKEN AT THE TIME INSTANTS t = n*Ts FOR n=0,1,...,2*N-1. THE
RESULTING SEQUENCE, x(n), IS INPUT OF A MOVING AVERAGE DIGITAL FILTER
WITH LENGTH L < N.   THE FILTER OUTPUT FOR n > N IS THE INPUT OF AN
IDEAL D/A CONVERTER WITH THE OUTPUT r(t).

                SPECIFY THE FOLLOWING PARAMETERS
               The value of amplitude A = 1
           Sinusoidal frequency F in hertz = 2
             Phase angle Gamma in degrees = 0
          Sampling interval Ts in seconds = 0.1
    Length of the moving average filter L = 3
                    Number of samples N = 21

> After you enter all the required data, the graph of $s(t)$ and $r(t)$ will appear
> in a MATLAB figure window as shown in Figure 9.2.

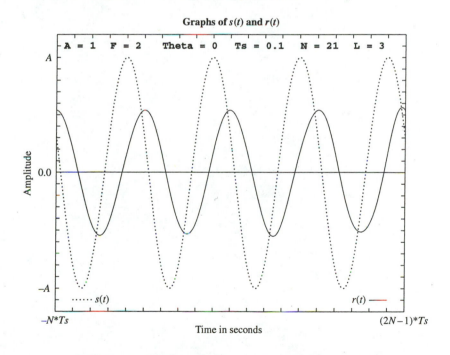

FIGURE 9.2   The plots of $s(t)$ and $r(t)$ used in Experiment 9

■ QUESTION 9.1   Use the graph shown in Figure 9.2 to determine the gain and phase-shift for the case illustrated.

Experiment 10

BASIC FILTER TYPES

BACKGROUND

One important application of discrete systems is filtering. In the design of a filter, the pole and zero locations are determined to obtain a desired gain and phase-shift as a function of the relative frequency, f. The gain and phase-shift can be specified only in the frequency range $0 \leq f \leq 1/2$. This is because the gain is an even function of f, the phase-shift is an odd function of f, and the system transfer function, $H(e^{j\omega})$, is a periodic function of the relative frequency, f, with a fundamental period equal to one.

■ QUESTION 10.1   Show that the gain is an even function and the phase-shift is an odd function of the relative frequency, f, by using the fact that the unit sample response, $h(n)$, is a real function of n.

■ QUESTION 10.2   Show that the system transfer function is a periodic function of the relative frequency, f, with a fundamental period of one. Discuss this result in relation to aliasing.

■ QUESTION 10.3   Use the results of Questions 10.1 and 10.2 to show that the transfer function can be specified only in the frequency range $0 \leq f \leq 1/2$.

EXPERIMENTAL PROCEDURE

There are several basic filter types upon which filter design is based. We shall examine some of these in this experiment. However, we first experimentally examine the gain and phase-shift of the moving average filter studied in the preceding experiment to fix the geometric ideas mentioned in the introduction to this chapter and to make use of the MATLAB function expr10 for this experiment.

The Moving Average (MA) Filter

We begin by examining the system in which, for the input $x(n)$, the response is the sum of the last two input values, so that

$$y(n) = x(n) + x(n - 1) \tag{10.1}$$

First, determine the system function, $H(z)$, of this system and specify the pole and zero locations. Then, run expr10 from the command window. You will be asked to specify the poles and zeros in polar form. The gain and phase-shift of the system with the system function given by eq. (10.2) are then computed:

$$H(z) = \frac{(z - z_1)(z - z_2) \cdots (z - z_m)}{(z - p_1)(z - p_2) \cdots (z - p_n)} \tag{10.2}$$

The data are available in the following forms:

- A screen plot of the z-plane poles and zeros
- Screen graphs of the gain and phase-shift

Using MATLAB's built-in line-editing capability, you can

- Add poles and/or zeros to those already entered
- Delete poles and/or zeros from those already entered
- Change the location of any poles and/or zeros already entered

This procedure is explained in the illustration section.

For the MA filter given by eq. (10.1), there is one zero on the unit circle at $z = -1 + j0$ and one pole at $z = 0$. Call for the screen plot of the z-plane poles and zeros to observe their locations. Modify their locations if they were not entered correctly. Now call for the screen graph of the gain and phase-shift. Use the geometric view of the gain and phase-shift discussed in the introduction to this chapter to verify the shape of the gain and phase-shift curves. Also compute the gain and phase-shift geometrically at $f = 0$, 0.25, and 0.50 cycles per sample. Verify your calculations from the gain and phase-shift graphs.

We now reexamine the MA filter studied in the preceding experiment. For the input, $x(n)$, the output is

$$y(n) = \frac{1}{L} \sum_{k=0}^{L-1} x(n - k) \tag{10.3}$$

Determine the system function, $H(z)$, of this filter and express the locations of all poles and zeros in polar form; also determine at what frequencies the gain is equal to zero.

Make a rough sketch of the gain and phase-shift curves for the case $L = 4$ by using the geometric view discussed in the introduction to Chapter III. Run expr10 for this case and compare the shape of the gain and phase-shift graphs with your sketch. Do the maxima of the gain curve lie halfway between the minima of the gain curve and do all the maxima have the same value? Discuss these results in terms of the geometric view. For this analysis, the following formula is useful:

$$\sum_{n=0}^{N-1} c^n = \begin{cases} N & \text{for } c = 1 \\ \dfrac{1 - c^N}{1 - c} & \text{for } c \neq 1 \end{cases}$$

Now analyze the case for $L = 8$ in the same manner as for $L = 4$.

We now shall use the MATLAB function expr10 to study some basic filter types upon which filter design is based.

The Notch Filter

The notch filter is used to eliminate a single frequency, f_1, by forming a notch in the gain curve. For $f_1 \neq 0$ or $1/2$, the filter has a pair of conjugate zeros and a pair of conjugate poles. Both zeros are at a radius of one, with one zero at an angle of $360 \cdot f_1$ degrees and the other zero at an angle of $-360 \cdot f_1$ degrees in which f_1 is the relative frequency (in cycles per sample) to be eliminated. The two poles are at the same angles as the zeros but at a radius of r_1. The radius r_1 determines the bandwidth of the notch filter.

■ QUESTION 10.4   Determine a difference equation relating the notch filter input, $x(n)$, and output, $y(n)$.

Use the geometric view to obtain an approximate expression for the gain at frequencies f close to f_1 as a function of only $(1 - r_1)/\Delta f$ in which $\Delta f = |f - f_1|$. With this result, determine the half-power frequencies as a function of r_1. Now run the MATLAB function and compare the experimental results with your calculations. Specifically compare and discuss the 3-db bandwidth versus r_1 and the shape of the gain curve in the frequency range about f_1. Also, experimentally explore the effect of f_1 on these results.

■ QUESTION 10.5   How would your results change if the conjugate pole-zero pair at the angle $-360 \cdot f_1$ degrees were eliminated? Why? Verify experimentally. What would be the difference equation relating the input and output of such a filter? Use this result to discuss the need for poles and zeros always to be in conjugate pairs.

The Band-Pass Filter

We examine only the simplest band-pass filter in this experiment. The filter we examine has a second-order zero at $z = 0$ and a pair of conjugate poles; both poles are at a radius of r_1, with one at an angle of $360 \cdot f_1$ degrees and the other at an angle of $-360 \cdot f_1$ degrees. Let f_m be the frequency at which the gain is a maximum.

■ **QUESTION 10.6**   Determine a difference equation relating the filter input, $x(n)$, and output, $y(n)$.

First choose $f_1 = 0.25$ and estimate the 3-db bandwidth versus r_1 by using the geometric view. Run `expr10` and compare the experimental results with your calculations. Note for this case that $f_m = f_1$, the gain curve is symmetric, and the phase-shift curve is antisymmetric about f_m.

Now choose $f_1 = 0.125$ and experimentally determine the upper and lower 3-db frequencies versus r_1. Compare the graph of the 3-db bandwidth versus r_1 with that obtained previously. Use the geometric view to also discuss why, in the present case, the gain curve is not symmetric and the phase-shift curve is not antisymmetric about f_m.

To further examine the asymmetry of the gain and phase-shift curves about f_m, experimentally study the frequency difference $\delta f = f_1 - f_m$ and the difference between the upper and lower 3-db frequencies as a function of f_1 and r_1.

■ **QUESTION 10.7**   Discuss the change in the filter characteristics studied above if each pole is replaced by a zero and each zero is replaced by a pole.

The All-Pass Filter

The name of the all-pass filter derives from its property that the filter gain is a constant so that it is the same at all frequencies. A trivial example of an all-pass filter is a delay, which has the system function $H(z) = z^{-k}$. The gain of this system is one at all frequencies and the phase-shift is $-k\omega$ so that the phase-shift varies linearly with frequency. The response of this system is simply its input delayed by k samples. In general, if a pole of an all-pass filter is located at $re^{j\theta}$, then there is a zero located at $\frac{1}{r}e^{j\theta}$. That is, each pole of an all-pass filter is matched by a zero located at the reciprocal conjugate of the pole position. Thus, the system function of an all-pass filter (except for a delay) is of the form

$$H(z) = C\frac{(z - z_1)(z - z_2)\cdots(z - z_k)}{(z - p_1)(z - p_2)\cdots(z - p_k)} \tag{10.4}$$

in which $z_n = 1/p_n^*$ and * indicates the conjugate.

■ QUESTION 10.8   Show that the system function of an all-pass filter also can be expressed in the form

$$H(z) = \frac{a_k z^k + a_{k-1} z^{k-1} + \cdots + a_1 z + 1}{z^k + a_1 z^{k-1} + \cdots + a_{k-1} z + a_k} \tag{10.5}$$

and that the coefficients are real if, for each pole of $H(z)$, there is a pole at the conjugate position. Use this result to determine the general form of a difference equation of an all-pass filter.

First, consider an all-pass filter with only one pole at $z = r$. Choose $r = 0.8$ and use expr10 to observe the general shape of the phase-shift curve and that the gain is a constant. Note that the scale but not the general shape of the phase-shift curve varies with r. To quantify the scale as a function of r, determine and plot a curve of the frequency at which the phase-shift is 90 degrees versus r for $0 < r < 1$.

Now consider an all-pass filter with two poles: one at $z = 0 + jr$ and the other at the conjugate location. Let f_+ be the frequency at which the phase-shift is 90 degrees and let f_- be the frequency at which the phase-shift is -90 degrees. Define the frequency $f_0 = (f_+ + f_-)/2$. The frequency f_0 is the frequency halfway between f_+ and f_-. How does f_0 compare with the frequency at which the phase-shift is 180 degrees? Determine and plot a curve of $(f_+ - f_0)$ versus r for $0 < r < 1$. Compare this curve with the one-pole case obtained above.

To determine the effect of the pole and zero angles, choose an all-pass filter with two poles: one at a radius of r and an angle of 45 degrees and the other at the conjugate location. Determine and plot a curve of $(f_+ - f_0)$ versus r for $0 < r < 1$ and compare this curve with that obtained above for the two-pole case. How does f_0 compare with the frequency at which the phase-shift is 180 degrees? Use the geometric view to discuss any differences between the phase-shift of the two-pole all-pass filters.

■ QUESTION 10.9   How would the phase-shift of an all-pass filter with two poles, one at a radius of r and an angle of 135 degrees and the other at the conjugate location, differ from the one just studied above?

Finally, examine the phase-shift of an all-pass filter with several poles. For this, begin with an all-pass filter with eight poles, each at a radius of $r = 0.8$ with angles 0, 45, -45, 90, -90, 135, -135, and 180 degrees. Theoretically determine the gain and phase-shift of this filter using the results obtained above. For your theoretical determination, note that this filter can be considered to be the tandem connection of the all-pass filters considered above. Now compare your theoretical expectation with that obtained experimentally.

Often, it is desired that a given system not introduce any phase distortion. For this, it is required that the phase-shift within the system band-pass varies linearly with frequency (since a phase-shift of $-k\omega$ corresponds to a simple delay of k samples). An all-pass filter that is connected in tandem with the given system can effect the desired result. The overall phase-shift of the tandem connection is $\theta_s(\omega) + \theta_a(\omega)$, in which $\theta_s(\omega)$ is the given system phase-shift and $\theta_a(\omega)$ is the phase-shift of the all-pass filter. The technique used is to determine the poles of the all-pass filter so that

$$\frac{d}{d\omega}[\theta_s(\omega) + \theta_a(\omega)] \approx \text{constant}$$

within the system band-pass. This is accomplished using the basic insight obtained in this experiment to determine the approximate required pole locations of the all-pass filter together with a computer to obtain the exact resulting phase-shift of the tandem connection.

AN ILLUSTRATION

The first MATLAB script that appears when you run `expr10` is the informational one shown below.

```
                    EXPERIMENT 10

   THIS PROGRAM WILL DETERMINE THE GAIN AND THE PHASE-SHIFT OF A STABLE LSI
DISCRETE SYSTEM FROM THE GIVEN POLES AND ZEROS OF THE SYSTEM FUNCTION, H(z)

YOU WILL BE ASKED TO SPECIFY:
   - ZERO locations in POLAR form,
   - POLE locations in POLAR form.

*** PRESS ENTER TO CONTINUE ***
```

After you press Enter, the next screen obtained is the one for entering the zero and pole locations. You enter these locations in polar form as a matrix input to MATLAB. The entered quantities must be enclosed in square brackets. The number of rows corresponds to the number of zero or pole locations. In each row, the radial distance is entered in the first column, and the angle (in degrees) is entered in the second column separated by a comma, whereas rows are separated by semicolons. This is shown below for $z_1 = 1$, $z_2 = -1$, $p_1 = 0.8j$, and $p_2 = -0.8j$.

```
ENTER POLAR FORM POSITIONS AS A MATRIX: [RADIAL DISTANCE, ANGLE(deg); ...]

Zeros = [1,0;1,180]
Poles = [0.8,90;0.8,-90]
```

To alter pole and zero values from the previous ones, press the up-arrow ↑ key repeatedly to recall these values and then make the necessary changes, including adding new values or deleting old ones.

After you enter all the data, graphs of the pole-zero plot and the magnitude-phase plot are shown in MATLAB figure windows. These graphs for the given locations are shown in Figures 10.1 and 10.2, respectively.

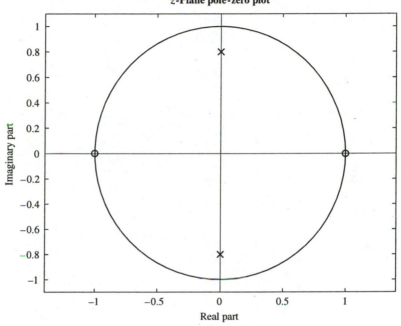

FIGURE 10.1 The pole-zero plot in Experiment 10

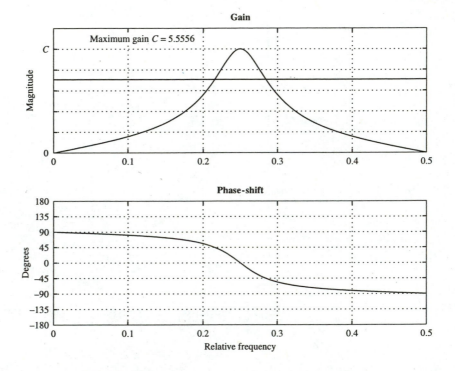

FIGURE 10.2   The magnitude-phase plot in Experiment 10

Experiment 11

FILTER TRANSIENT RESPONSE

BACKGROUND

Each of the basic filter types investigated in Experiment 10 has a desirable gain and/or phase-shift characteristic that is used in filter design. However, the gain and phase-shift are defined only for sinusoidal inputs. What is the response of these filters to other inputs? For example, the notch filter has zero gain at frequency, f_1, so the filter response is zero to an input sinusoid with frequency f_1. But what is the response of this filter to a sinusoidal pulse with frequency f_1? This question involves an investigation of the filter transient response. In this experiment, we shall investigate the transient response of the basic filter types studied in the last experiment.

EXPERIMENTAL PROCEDURE

The MATLAB function `expr11` allows us to study a discrete filter that can be described by a difference equation with the input, $x(n)$, and the output, $y(n)$, given by

$$y(n) = b_0 x(n) + b_1 x(n-1) + \cdots + b_M x(n-M)$$

$$+ a_1 y(n-1) + a_2 y(n-2) + \cdots + a_N y(n-N)$$

You will be asked to specify the coefficients of the difference equation. Then you can choose one of six different inputs, $x(n)$:

1. $x(n) = \delta(n)$ in which $\delta(n) = \begin{cases} 1 & \text{for} \quad n = 0 \\ 0 & \text{for} \quad n \neq 0 \end{cases}$ is the unit sample

2. $x(n) = u(n)$ in which $u(n) = \begin{cases} 1 & \text{for} \quad n \geq 0 \\ 0 & \text{for} \quad n < 0 \end{cases}$ is the unit step

3. $x(n) = \cos(\omega n + \theta)\, u(n)$

4. $x(n) = p(n, w)$ in which $p(n, w) = \begin{cases} 1 & \text{for } 0 \leq n \leq w \\ 0 & \text{otherwise} \end{cases}$ is a rectangular pulse

5. $x(n) = \cos(\omega n + \theta)\, p(n, w)$ is a sinusoidal pulse

6. $x(n) =$ an input of your choice

The output will be computed for $0 \leq n \leq 1000$. The graphs of $x(n)$ and $y(n)$ for $0 \leq n \leq 100$ are shown in the MATLAB figure window. The abscissa is marked at the sample numbers 0 and 100. The input, $x(n)$, is plotted with a scale of 0.1 unit per division. However, the graph of the output, $y(n)$, is scaled so that the maximum amplitude is ten divisions; the scale is printed at the top of the graph. After you click the left mouse button in this window, a menu is presented to choose a new display interval or to terminate the display. To display new plotted values, you can choose $0 \leq N_1 \leq n \leq N_2 \leq 1000$ for zooming-in or zooming-out of the graph that is displayed. For some filters and inputs, the first few values of $y(n)$ are much larger than succeeding values. In such a case, the latter values appear quite small on the graph since the scale is determined by the first few values. For such cases, you can examine these values with greater precision by a proper choice of N_1 because the scale is then determined by the plotted values of $y(n)$.

The transient behavior of a large class of discrete systems can be investigated using this experiment. We begin by investigating the transient behavior of the basic filter types studied in the last experiment.

The Notch Filter

As discussed in the last experiment, the notch filter is used to eliminate a single frequency by forming a notch in the gain curve. The filter has a pair of conjugate zeros and a pair of conjugate poles. Both zeros are at a radius of one, with one zero at an angle of $360 \cdot f_1$ degrees and the other zero at an angle of $-360 \cdot f_1$ degrees in which f_1 is the relative frequency (in cycles per sample) to be eliminated. The two poles are at the same angles as the zeros but at a radius of r_1. The radius r_1 determines the bandwidth of the notch filter.

First, determine a difference equation relating the notch filter input, $x(n)$, and output, $y(n)$. Choose the notch frequency, f_1, to be 0.25 cycles per sample and $r_1 = 0.9$. Use the program to determine the filter response to input #3 listed above with a frequency of f_1 cycles per sample and a phase of zero. Is it obvious from the transient response that the gain at this frequency is zero? Now run `expr11` with various values of r_1 in the range $0 \leq r_1 < 1$. Describe how the transient response changes as a function of r_1.

Now choose Input 5 listed above with $f_1 = 0.25$ and study the transient response as a function of the pulse width, w, and the value of r_1 of the notch filter. Specifically, discuss the source of the two output pulses. Plot the fall times of the pulses as a function of r_1 and compare them with that obtained with Input 4, the rectangular pulse.

Finally, examine the unit sample response of the notch filter as a function of r_1. From this response, can you determine why the output of a notch filter is zero for an input sinusoid of frequency f_1?

The Band-Pass Filter

The band-pass filter examined in the last experiment is the simplest band-pass filter. The filter has a second-order zero at $z = 0$ and a pair of conjugate poles; both poles are at a radius of r_1 with one at an angle of $360 \cdot f_1$ degrees and the other at an angle of $-360 \cdot f_1$ degrees. The filter bandwidth and the center frequency of the band-pass region, f_c, are controlled by f_1 and r_1.

First determine a difference equation relating the band-pass filter input, $x(n)$, and output, $y(n)$. Choose $f_1 = 0.25$ cycles per sample and $r_1 = 0.9$. Use the program to determine the filter response to Input 3 listed above with a frequency of 0.25 cycle per sample. Examine the effect of the phase angle, θ, on the transient response. Is there a value of θ for which there is no transient? Now run `expr11` with various values of r_1 in the range $0 < r_1 < 1$. Describe how the transient response changes as a function of r_1 and θ. Plot a graph of the rise time versus r_1. Can the gain and phase-shift of the filter be used to predict any aspect of the output? Why?

Now examine the response of the band-pass filter with $f_1 = 0.25$ and $r_1 = 0.9$ to the sinusoidal pulse, Input 5, as a function of f, θ, and w. For $n < w$, the response should be identical to that obtained with Input 3. Why? Examine and discuss the response for $n > w$. Are there values of f, θ, and w for which $y(n) \approx 0$ for $n > w$?

Finally, examine $h(n)$, the unit sample response of the band-pass filter. Can you determine from $h(n)$ why the filter is a band-pass filter ?

The All-Pass Filter

The name of the all-pass filter derives from its property that the filter gain is a constant so that it is the same at all frequencies. In general, as we discussed in the last experiment, each pole of an all-pass filter is matched by a zero located at the reciprocal conjugate of the pole position. Also, its system function can be expressed in the form

$$H(z) = \frac{a_k z^k + a_{k-1} z^{k-1} + \cdots + a_1 z + 1}{z^k + a_1 z^{k-1} + \cdots + a_{k-1} z + a_k} \qquad (11.1)$$

in which the coefficients are real since $h(n)$ is a real function of n. Use this result to determine the general form of a difference equation of an all-pass filter.

First consider an all-pass filter with only one pole at $z = r$ and one zero at $z = 1/r$. Choose $r = 0.8$ and use the program to observe the filter response to Input 3, the sinusoidal step. The output, $y(n)$, approaches steady-state as $n \rightarrow \infty$. Examine $y(n)$ for large values of n as a function of the input frequency, f, with the input phase, θ, equal to zero. Note that in steady-state, the maximum amplitude of $y(n)$ varies with f. Does this mean that the filter implemented is not an all-pass filter? You must, of course, check the implemented filter. If you are certain it is an all-pass filter, what could be the source of this result? Clearly, there is only one parameter available that could be the source of the problem and that is the input phase angle, θ. Check to determine whether this is the source of the problem. If indeed you determine that it is, explain why θ has the observed effect on the steady-state behavior because, after all, shouldn't the steady-state behavior be independent of the input phase?

With the sinusoidal step, Input 3, examine the transient component of $y(n)$ as a function of the filter parameter, r, and compare it with the step, Input 2. To further examine the transient response, choose the sinusoidal pulse, Input 5, and observe $y(n)$ for $n > w$. One interesting case is for $f = 0.49$. Discuss $y(n)$ for this case as a function of the input phase, θ.

Now consider an all-pass filter with two poles: one at $z = 0 + jr$ and the other at the conjugate location. The zeros positions, of course, are at the reciprocal conjugate of the pole positions. Repeat your analysis above for this filter. Discuss the similarities and differences of the transient responses for these two cases.

AN ILLUSTRATION

■

The MATLAB script that appears when you run `expr11` is shown below.

EXPERIMENT 11

THE DIFFERENCE EQUATION RELATING THE FILTER
    INPUT, x(n), AND OUTPUT, y(n), IS:

```
y(n) = b0*x(n) + b1*x(n-1) + ... + bM*x(n-M)
        + a1*y(n-1) + a2*y(n-2) + ... + aN*y(n-N)
```

SPECIFY THE DESIRED VALUES OF THE FILTER COEFFICIENTS AS VECTORS,
e.g.,
        [b0, b1, ..., bM] as a vector b
and   [a1, a2, ..., aN] as a vector a

b =

a =

The filter coefficients are entered as vectors enclosed by square brackets with the values separated by commas. After you enter the coefficients, a MATLAB menu window shown below will appear from which the desired filter input can be chosen.

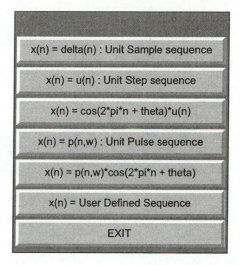

After you choose the filter input, the MATLAB command window for choosing the desired parameters of the input signal appears. For example, if you choose Input 5, the script shown below will appear.

```
YOU HAVE CHOSEN THE SINUSOIDAL PULSE SEQUENCE

        x(n) = cos(2*pi*f*n + theta)p(n,w)

normalized frequency f =
phase angle in degrees =
pulse width in samples =
```

Once the parameters of the input sequence are chosen, a MATLAB figure window will appear, displaying $x(n)$ and $y(n)$. After you click the left mouse button, the following MATLAB menu window will appear.

Zoom the n axis

DONE

If you choose the zoom option, then the following script will appear.

```
THE SEQUENCE WILL BE PLOTTED FOR 0 <= N1 <= n <= N2 <= 1000

Specify the value of N1 =
Specify the value of N2 =
```

After you supply the necessary values, another MATLAB figure window will display the input and the output sequences over the chosen interval with proper scaling.

For example, if the coefficients of the chosen filter have the values $b_0 = 1$, $a_1 = 0.9$, and all the others are zero, and if you choose Input 5 with the parameter values $f = 0.05$ sample per cycle, $\theta = 0$ degrees, and pulse width $w = 16$ samples, then the screen graph shown in Figure 11.1 appears. Then, if you choose the values $N_1 = 0$ and $N_2 = 50$ for display, the screen graph shown in Figure 11.2 appears. Choosing Exit from the signals menu terminates the program.

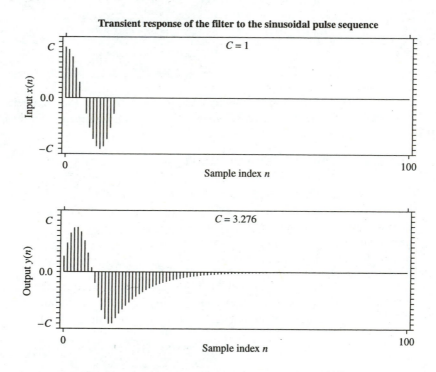

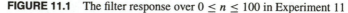

FIGURE 11.1   The filter response over $0 \le n \le 100$ in Experiment 11

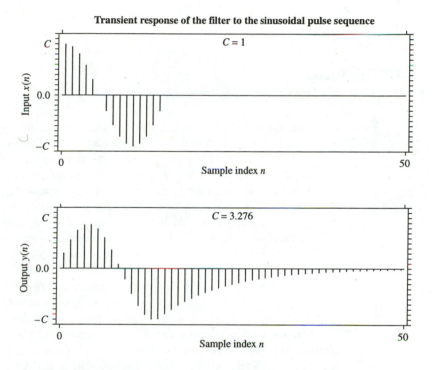

Transient response of the filter to the sinusoidal pulse sequence

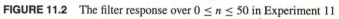

FIGURE 11.2   The filter response over $0 \le n \le 50$ in Experiment 11

BUTTERWORTH FILTER DESIGN

BACKGROUND

The ideas studied in Experiment 10 about the simple band-pass filter will be used to design a second-order Butterworth filter. This is a band-pass filter for which the graph of the gain versus frequency has a shape that is closer to the ideal rectangle than that of the simple band-pass filter studied in Experiment 10. Extensive studies of the discrete Butterworth filter have been made and formulas for its design have been developed.[1] The objective of this experiment is not to verify the known formulas but to use the geometric view discussed in the introduction to Chapter III to determine the direction in which poles and/or zeros should be moved to obtain a gain characteristic closer to the desired characteristic.

The procedure used in this experiment is an iterative one in which each successive movement of the poles and/or zeros brings the curve of the gain versus frequency closer to the desired one. In this design technique, no consideration is initially given to obtaining a desired phase-shift characteristic. Once the poles and zeros of a filter with the desired gain characteristic are determined, an all-pass filter is designed so that the phase-shift of the determined filter plus that of the all-pass filter is equal to the desired phase-shift. The tandem connection of the determined filter and the all-pass filter then has the desired gain and phase-shift characteristic. Of course, each of the two filters is not synthesized and connected in tandem to obtain the desired filter. Rather, the difference equation relating the input and output of the tandem connection is determined. This

---

[1] See, for example, D. J. DeFatta, J. G. Lucas, and W. S. Hodgkiss, *Digital Signal Processing: A System Design Approach*, John Wiley & Sons, Inc., New York, NY, 1988, and M. T. Jong, *Methods of Discrete Signal and System Analysis*, McGraw-Hill Book Co., New York, NY, 1982.

difference equation is then realized. For this experiment, only a filter with a certain desired gain characteristic will be determined. For those interested in going further, an all-pass filter can be determined so that the phase-shift of the desired filter varies linearly (approximately) with frequency within the band-pass region. Such a filter would have the desired gain characteristic with minimal phase distortion.

EXPERIMENTAL PROCEDURE

PART 1

First choose the system to have four zeros at the origin and choose the system to have four poles: one at a radius of 0.8 and an angle of $90 - \phi$ degrees, one at a radius of 0.8 and an angle of $90 + \phi$ degrees, and the other two poles located at the conjugate positions of the two given poles. Choose $\phi = 25$ degrees and note that the gain curve for this case has a dip at its center, $f = 0.25$. Now observe the gain curve as ϕ is decreased. Note that at its center, $f = 0.25$, the curve goes from convex downward to convex upward. Mathematically, the second derivative of the curve with respect to frequency at $f = 0.25$ goes from being positive to being negative. Thus there is a value of ϕ for which the second derivative at this point is zero. Determine this value of ϕ experimentally. For this value of ϕ, the gain characteristic is said to be maximally flat. This filter is a second-order Butterworth filter with a center frequency of 0.25 cycle per sample. What is the 3-db bandwidth of this filter? Determine a difference equation relating the input and output of the Butterworth filter determined. Use the program for Experiment 11 to examine the filter unit sample response and its transient response to various input sequences.

PART 2

The filter just determined was simple because all the poles have the same radius and there was only one parameter, ϕ, to vary to obtain the maximally flat condition. To complicate matters a bit, a second-order Butterworth filter with a different center frequency will be determined. Again choose the system to have four zeros at the origin and choose the system to have four poles: one at a radius of 0.8 and an angle of 35 degrees, one at a radius of 0.8 and an angle of 55 degrees, and the other two poles located at the conjugate positions of the two given poles. Note that the gain curve is not maximally flat. Obtain the maximally flat gain characteristic by moving the pole at 55 degrees and its conjugate. Note that both the radius and its angle must be changed. The best iterative procedure is to vary only one parameter at time. Using the geometric view, discuss your adjustment procedure and your reason for each adjustment made. Also use the geometric view to explain why the maximally flat condition is not attained with all poles at the same radial distance as was the case in the previously determined Butterworth filter. What are the bandwidth and the center frequency of the Butterworth filter you have determined? Determine a difference equation relating the input and output of the Butterworth filter determined. Use the program for Experiment 11 to examine the filter unit-sample response and its transient response to various input sequences.

PART 3

The Butterworth filter characteristic is an approximation to the ideal band-pass characteristic, which is rectangular. The Butterworth filter designed is called a second-order Butterworth because it has two poles in the upper half of the z-plane. If n poles are in the upper half of the z-plane and the gain characteristic is maximally flat, the filter is called an nth-order Butterworth. However, it would seem that a second-order Butterworth characteristic can be improved by adding zeros on the unit circle at the band edge. If a zero were placed at a radius of 1 and an angle of ψ degrees, then the gain must be zero at a relative frequency of $180 \cdot \psi$ cycles per sample. Use the geometric view to explain why.

Consider the Butterworth filter determined in Part 1 and add four zeros all at a radius of 1 with angles 55, -55, 125, and -125 degrees. Examine the gain and phase-shift characteristic of the resulting filter. Note that the gain characteristic is no longer maximally flat. Now keep the eight zeros in place and change the pole angles, keeping their radii at 0.8 and the poles symmetrically placed relative to the z-plane ordinate so that the gain and phase-shift curves are symmetric about $f = 0.25$ cycle per sample. Note the effect on the gain and phase-shift characteristic. Is there a position where the gain characteristic is again maximally flat? How does the band-pass bandwidth compare with that of the Butterworth filter designed in Part 1? Compare the rate of gain decrease outside the band-pass region with that of the Butterworth filter. Which filter has the greater phase distortion?

Try to reduce the phase distortion of the filter determined above by changing the pole angles, keeping their radii at 0.8, and keeping the poles symmetrically placed relative to the z-plane ordinate as above. Study how the gain and the phase-shift change as the pole angles are changed. Discuss the trade-off between the desired rectangular gain curve and the phase distortion.

It is seen from the experimental design procedure in this experiment why the iterative technique can be difficult, especially for filters with several poles. The approximate number of poles and zeros with their locations can be determined using the geometric view. However, you observed in your determination of the last Butterworth filter that the gain characteristic is somewhat sensitive to the pole locations. This sensitivity is greater for filters with a larger number of poles. This, then, is the place of formulas in the design of high-order filters. However, slight modifications from the standard design, as in this experiment, can be achieved using the geometric view and the iterative technique used in this experiment.

AN ILLUSTRATION

See the screen illustration for Experiment 10 since the screens obtained in this experiment are identical.

CHAPTER IV

QUANTIZATION

Computer quantization is an important aspect of discrete systems and digital signal processing because it limits the accuracy with which the computer can process data. We include a basic discussion of quantization in this introduction and in each of the experiments in this chapter because many textbooks on discrete systems contain only a cursory discussion of the effect of quantization on system performance.

As we communicate using words, so does the computer. The words used by almost all digital computers are binary, which are a special type composed of only two characters, 1 and 0. Thus, one possible binary word is 10010111. Each character is called a bit so the word illustrated has a length of eight bits and is called an eight-bit word. (An eight-bit word is often called a *byte*.) The number of different words available is equal to the number of possible different sequences of 1's and 0's. For an eight-bit word, this number is $2^8 = 256$. In general, there are 2^b different b-bit words available so any number processed by the computer first must be quantized into one of 2^b different values. This can result in a quantization error because the number to be processed may not be exactly equal to one of the 2^b available values. The specific value of b used in any particular application depends not only on the design of the computer, but also on the data type — such as single- or double-precision variables — being used. There are three main types of quantization errors of interest in discrete systems: *coefficient* quantization error, *A/D* quantization error, and *multiplication* quantization error.

A given discrete system is realized on a computer by relating the input and output by means of a difference equation. Coefficient quantization error is an error that occurs when the value of each coefficient of the difference equation is quantized into one of the 2^b available values before it is stored. The coefficients of the discrete system implemented on the computer and those of the desired discrete system then differ due to the coefficient quantization. The resulting difference between the characteristics of the implemented system and those of the

desired system can be analyzed theoretically simply by determining characteristics of the implemented system using the quantized coefficient values.[2] This
difference won't be analyzed in Chapter IV because it is fixed and easily analyzed theoretically.

A/D quantization error occurs because each sampled value of an analog
waveform must be quantized before it is stored in the computer as one of the
2^b available values. To study this error, which varies from sample to sample, we
will establish a model of the error.

Multiplication quantization error is the quantization error introduced in the
output of a discrete system that results from the multiplication of sequences
within the system by the difference equation coefficients. Even though the coefficients and sequence values are quantized, their product may not be one of the
2^b available values and so must be quantized before being stored. These multiplication quantization errors circulate within the discrete system and can result
in output errors many times larger than their original values. We will also study
this error in Chapter IV by establishing a model of the error.

The reason for establishing a model for the A/D and multiplication quantization errors is to avoid the difficult nonlinear analysis of the quantizer. Also, it
often is sufficient to know only certain statistical properties of the error. Thus,
the models of the error we will establish will be what we call statistically equivalent models. We define two sequences to be statistically equivalent if they have
the same statistical properties. For example, the values of a random sequence
generated on your computer may well differ from those of a random sequence
generated on another computer. However, two sequences are statistically equivalent if they have the same statistical properties. It is easy to prove the following
result:

> Let the response of a given linear shift-invariant (LSI) system to
> $x_1(n)$ be $y_1(n)$ and the system response to $x_2(n)$ be $y_2(n)$. Then,
> $y_1(n)$ and $y_2(n)$ are statistically equivalent if $x_1(n)$ and $x_2(n)$ are
> statistically equivalent.

This result will enable us to study the statistical effects of the error on the
system response by establishing a model of the actual error that has only the
same statistical properties as the actual error. With such a statistically equivalent
model, we can avoid the nonlinear analysis of the quantizer in determining the
statistical properties of the exact error.

---

[2]See Section 7.6 of J. G. Proakis and D. G. Manolakis, *Digital Signal Processing*, Third Edition,
Prentice Hall, Inc., Englewood Cliffs, NJ, 1996, for a discussion of the sensitivity of filter characteristics to coefficient values.

Experiment 13

A/D QUANTIZATION ERROR

BACKGROUND

As discussed in the introduction to Chapter I, a general model of an A/D converter is the tandem connection of a sampler and a quantizer as shown in Figure 13.1. The quantization error was essentially zero in the sampling experiments of Chapter I. In effect, the computer was modeled as one with infinite word length ($b = \infty$). This was done in order to study the effects of sampling without any complicating effects due to quantization. In this experiment, the effect of a finite word length in A/D conversion will be examined.

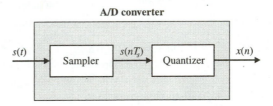

FIGURE 13.1   Analog-to-digital converter

Since a computer using b-bit words has only 2^b different words available, each sample value of $s(t)$ is quantized into one of 2^b different values. The A/D quantization error, $q(n) = x(n) - s(nT_s)$, is the difference between the quantized value, $x(n)$, and the sample value, $s(nT_s)$. The A/D converter output thus can be expressed as

$$x(n) = s(nT_s) + q(n) \tag{13.1}$$

The sequence $x(n)$ is then processed by a discrete system and a D/A converter. Because we are concerned with only linear discrete systems in this text, we can use superposition to express the discrete system response, $z(n)$, as

$$z(n) = y(n) + e(n) \tag{13.2}$$

in which $y(n)$ is the discrete system response to $s(nT_s)$ and $e(n)$ is the discrete system response to $q(n)$. Also, since D/A converters generally are linear systems, we again can use superposition to express the D/A output, $r(t)$, as

$$r(t) = y_a(t) + e_a(t) \tag{13.3}$$

in which $y_a(t)$ is the analog D/A output due to $y(n)$ and $e_a(t)$ is the analog D/A output due to $e(n)$.

■ QUESTION 13.1   Discuss why a D/A converter is generally designed to be a linear system. For this, it would be helpful to consider the consequences of it not being a linear system.

We thus note that, because the discrete system and the D/A converter are linear systems, the tandem connection is a linear system so that superposition can be used. Consequently, the error $e_a(t)$ is due to only the error sequence $e(n)$ which, in turn, is due to only the error sequence $q(n)$ as shown in Figure 13.2. We thus could study the errors introduced by A/D quantization directly by generating the sequence $q(n)$ and using it as the input of the linear discrete system. The system output then would be the output error due to A/D quantization.

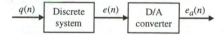

FIGURE 13.2   System model for the analog output error due to A/D quantization

■ QUESTION 13.2   The preceding argument depends critically on the linearity of the discrete system. Examine how the argument fails if the discrete system is not linear by discussing the case of a discrete system for which $z(n) = x^2(n)$.

The larger the computer word length, the more different values are available into which the sample values of $s(t)$ can be quantized. Thus, it is clear that the larger the word length, the smaller can be $q(n)$ and consequently $e(n)$. But it generally is desirable to use the smallest word length necessary because a smaller word length requires less computer memory and the data processing will be faster. Thus, in the design of a processor, it is desirable to know the shortest word length required to achieve an acceptably small error. Also, if a commercially available processor is to be chosen, then it is necessary to know whether the processor word length is sufficiently large.

To determine $e(n)$, first $q(n)$ must be known. But the specific sequence $q(n)$ depends on the particular waveform sampled. Thus $s(t)$ would have to be known before $q(n)$ could be known. But $s(t)$ often is not known a priori especially if it is an information bearing signal. However, it often is not necessary to know the specific sequence, $e(n)$. Rather, only certain statistical properties of $e(n)$ are required. The approach we shall take is to determine the statistical properties of $e(n)$ without knowing $q(n)$ specifically. By asking for only statistical information, we shall see that, under certain conditions, it will not be necessary to know $s(t)$. That is, under certain conditions that we shall investigate, all the necessary statistical information of $e(n)$ can be determined without specifying $s(t)$. For such cases, the statistical properties of $e(n)$ are independent of the input signal, $s(t)$, and the minimum word length for which some measure of the error (such as the mean-square error or the peak value of the error) is less than some specified value can be determined without any specification of $s(t)$.

The approach we use is based on statistical equivalence as discussed in the introduction to Chapter IV. Two different sequences that have the same statistical properties are said to be statistically equivalent. Clearly, the statistical properties of $e(n)$ can be determined by examining a sequence that is only statistically equivalent to $e(n)$. Let the response of the discrete system in Figure 13.2 be $e_1(n)$ to the input $q_1(n)$ and be $e_2(n)$ to the input $q_2(n)$. Then, from the result stated in the introduction to Chapter IV, $e_1(n)$ and $e_2(n)$ are statistically equivalent if the inputs $q_1(n)$ and $q_2(n)$ are statistically equivalent. That is, two different LSI system responses are statistically equivalent if their corresponding inputs are statistically equivalent. Thus, the desired statistical properties of $e(n)$ can be determined by using an input sequence that may not be identical to the specific $q(n)$ obtained but only statistically equivalent to it. The purpose of this experiment is to study the statistical properties of $q(n)$ experimentally so that a statistically equivalent sequence can be generated. The statistically equivalent sequence then can be used to determine the statistical properties of $e(n)$ and $e_a(t)$ either theoretically or experimentally.

In A/D quantization, the amplitude range usually is divided into equal intervals $\tilde{q}$ units wide so that $x(n) = 0$ if

$$-\frac{1}{2}\tilde{q} \leq s(nT_s) < \frac{1}{2}\tilde{q} \tag{13.4}$$

and $x(n) = \tilde{q}$ if

$$\frac{1}{2}\tilde{q} \leq s(nT_s) < \frac{3}{2}\tilde{q} \tag{13.5}$$

Generally, $x(n) = k\tilde{q}$ for $k = 0, \pm 1, \pm 2, \pm 3$ if

$$\frac{2k-1}{2}\tilde{q} \leq s(nT_s) < \frac{2k+1}{2}\tilde{q} \tag{13.6}$$

Figure 13.3 is a graph of $x(n)$ versus $s(nT_s)$, which is the transfer characteristic of a quantizer. A quantizer with the transfer characteristic shown in Figure 13.3 is called a uniform quantizer.

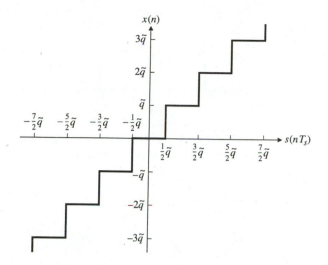

FIGURE 13.3   Transfer characteristic of a uniform quantizer

Observe that the values of the quantization error, $q(n)$, always lie between $-\frac{1}{2}\tilde{q}$ and $\frac{1}{2}\tilde{q}$. As an example, consider the case for which $\tilde{q} = 0.1$. If $s(nT_s) = 0.99$, we then have $x(n) = 1$ and $q(n) = -0.01$; if $s(nT_s) = 0.91$, then $x(n) = 0.9$ and $q(n) = +0.01$. However, note that $s(nT_s)$ need not be 0.91 for $q(n)$ to be 0.01 in our example; $q(n)$ also would be 0.01 if $s(nT_s)$ were equal to -0.19 or -0.09 or $+0.01$ or $+0.11$ or $+0.21$. Furthermore, if $s(nT_s)$ were equally likely to have any value in a 0.1 amplitude range, then $q(n)$ in our example would be equally likely to have any value between -0.05 and $+0.05$. Check these results by choosing the signal to be $s_c(t)$, a constant, and choose a small sample size. Every sample of the error will have the same value for this case so the graph obtained is independent of the sample size. Also, choose other values for the constant to make certain you understand the determination of the quantization error.

If the values of $q(n)$ are uniformly distributed so that they are equally likely to have any value in the interval between $-\frac{1}{2}\tilde{q}$ and $\frac{1}{2}\tilde{q}$, then a graph of $P_1(q)$, the probability density of the values of $q(n)$, is as shown in Figure 13.4a; Figure 13.4b is an equivalent normalized graph.

The probability that $0 \leq q(n) < 0.01\tilde{q}$ is just the area under the probability density graph in that interval. For our example shown in Figure 13.4, it is

$$\int_{0}^{0.01\tilde{q}} P_1(q)dq = \int_{0}^{0.01\tilde{q}} \frac{1}{\tilde{q}}\, dq = 0.01 \qquad (13.7)$$

The total area under any probability density graph equals one since it is a certainty that some event will occur. Thus, no matter what is the graph of $P_1(q)$,

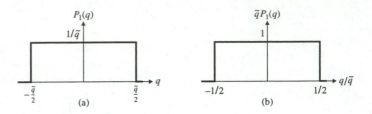

FIGURE 13.4   The probability density of a uniformly distributed quantization error: (a) density of $q(n)$; (b) normalized density

we must have

$$\int_{-0.5\tilde{q}}^{0.5\tilde{q}} P_1(q)dq = 1 \tag{13.8}$$

since it is a certainty that $q(n)$ will have some value in the interval $-\frac{1}{2}\tilde{q} \leq q < \frac{1}{2}\tilde{q}$.

The essential theoretical result that is the basis of the statistical model being developed in this experiment is as follows:

> The values of the sequence $q(n)$ are equally likely to have any value in the interval $-\frac{1}{2}\tilde{q} \leq q < \frac{1}{2}\tilde{q}$ and, for $n_1 \neq n_2$, the value of $q(n_1)$ is independent of the value of $q(n_2)$ if (1) the sampled sequence, $s(nT_s)$, is not periodic; (2) the probability that $s(mT_s) = s(nT_s)$ for $m \neq n$ is zero; and (3) the width of the quantization interval, $\tilde{q}$, is sufficiently small.

Note that under the conditions of this theoretical result, we can study the effects of the quantization error without concern for the specific waveform, $s(t)$, being sampled since the quantization error is independent of it. Also, we can use the system depicted in Figure 13.2 to analyze statistically the contribution of the quantization error to the system output without specific analysis of the nonlinear quantizer. A random sequence of length N with the statistical properties given above can easily be generated in MATLAB using the rand function as

```
q = q_tilde*(rand(N,1)-0.5);
```

An equivalent function is available in most higher-level languages. The objective of this experiment is to investigate this theoretical expectation. In particular, we shall investigate how small $\tilde{q}$ must be for the theoretical result to be valid.

EXPERIMENTAL PROCEDURE
■

The MATLAB function for this experiment is `expr13`. After running `expr13`, you will be given two input choices: the number of decimal places D and the number of samples N, and several choices for the signal $s(t)$. The chosen signal is sampled at times $t = nT_s = 0, 1, 2, 3, \ldots$, and N samples are used for quantization error analysis. The signal choices are

- $s_c(t) = $ a constant of your choice between -3.2 and 3.2
- $s_1(t) = \sin(t/11)$
- $s_2(t) = \frac{1}{2}\{\sin(t/11) + \cos(t/79)\}$
- $s_3(t) = \frac{1}{3}\{\sin(t/11) + \cos(t/79) + \sin(t/31)\}$
- $s_4(t) = \frac{1}{4}\{\sin(t/11) + \cos(t/79) + \sin(t/31) + \sin(t/57)\}$
- $s_r(t) = $ a pseudo-random sequence uniformly distributed between -1 and 1

The sample values are quantized to D decimal places. Choosing $D = 0$ results in $s(nT_s)$ being quantized to the nearest integer so that $\tilde{q} = 1.0$; generally, $\tilde{q} = 10^{-D}$. The graph obtained in the top half of the MATLAB figure window is a normalized bar-graph of the fraction of samples of $q(n)$ that lie in each $0.01\tilde{q}$-wide interval. For example, if the graph of the probability density were as shown in Figure 13.4, then the height of each bar would be 0.01.

However, the experimentally obtained bar graph can be expected to be in error. The error is due to the sample size. The sample size, N, is the number of values of $q(n)$ that are used to determine the screen graph. The larger the sample size, the smaller is the expected error. To understand this, consider determining the average exam grade of a large number of students. If the average exam grade is estimated by averaging two randomly chosen exam grades (a sample size of two), then the error of the estimate is expected to be large. Clearly, the larger the sample size used to estimate the average exam grade, the smaller is the expected error of the estimate. Similarly, in determining the bar-graph, the larger the sample size, N, the smaller is the expected error in the experimentally obtained bar graph. For this error to be acceptably small, a sample size of at least $20,000$ is required. Although a larger sample size is desirable, it requires more time to obtain the screen graph because the time required clearly increases linearly with the sample size.

First choose $s(t)$ to be the sinusoidal waveform $s_1(t)$ from the signal menu and obtain the probability bar graph of $q(n)$ for $D = 0, 1, 2, 3,$ and 4.

■ QUESTION 13.3   Show that the sequence $s_1(nT_s)$ is not periodic. It is helpful to show first that a sinusoidal sequence is periodic only if $f = F/F_s$ is a rational number. If f is a rational number, how can the fundamental period of the sinusoidal sequence and the number of cycles of $s(t)$ that must be sampled to obtain one fundamental period of the sequence be determined?

Since the sequence $s_1(n T_s)$ is not periodic, the samples are uniformly distributed over a fundamental period of $s_1(t)$. That is, the fraction of samples with values equal to $\sin \theta$ for $\alpha < \theta < \alpha + \delta$ is the same for *any* value of α.

■ QUESTION 13.4   The first bar to the right of $q = 0$ is the probability that $0 \leq q(n) < 0.01$. Compute this probability for $D = 0$ and compare your calculated value with that obtained experimentally. For this calculation, note that a sample value of the sinusoidal waveform $s_1(t)$ is equally likely to be taken at any angle of the sinusoid since $s(nT_s)$ is not periodic. Use this to show that the probability that $0 \leq q(n) < 0.01$ is equal to the fraction of a sinusoidal period that $\sin(\phi) > 0.99$ plus the fraction of a sinusoidal period that $-0.01 < \sin(\phi) \leq 0$, and so compute the desired probability.

■ QUESTION 13.5   Now compute the probability that $0.01 \leq q(n) < 0.02$ for $D = 0$ and compare your calculated values with those obtained experimentally.

■ QUESTION 13.6   Repeat the computation of the two probabilities above for $D = 1$ and compare your calculated values with those obtained experimentally.

■ QUESTION 13.7   Using your above calculations as a basis, discuss why the values of $q(n)$ become more uniformly distributed with increasing values of D.

■ QUESTION 13.8   Experimentally determine the values of D for which the values of $q(n)$ are uniformly distributed.

Now obtain the probability bar graphs of $q(n)$ using the signal $s_2(t)$ with $D = 0, 1, 2$, and 3. For what values of D are the values of $q(n)$ uniformly distributed? From this simple experiment, what value of D would you expect to be required for $q(n)$ to be uniformly distributed if the waveform, $s(t)$, is composed of the sum of a large number of nonharmonically related sinusoids? Discuss why. Check your expectation by obtaining the probability bar graphs using signals $s_3(t)$ and $s_4(t)$.

SAMPLE INDE-   You now have experimentally determined the maximum value of $\tilde{q}$ for which
PENDENCE   the values of $q(n)$ are uniformly distributed. However, even though the values of $q(n)$ are uniformly distributed, they may not be independent. It is now necessary to determine the maximum value of $\tilde{q}$ for which $q(n_1)$ is independent of $q(n_1)$ for $n_1 \neq n_2$. The probability bar graph of $e_2(n) = \frac{1}{2}[q(n) + q(n-1)]$ will be used for this determination. This bar graph is plotted in the lower half of the MATLAB window.

First, the expected probability density distribution of $e_2(n)$ must be determined. For this, we note that $e_2(n) = e$ when $q(n) + q(n-1) = 2e$. This is so when, for any value of q, we obtain $q(n) = q$ and $q(n-1) = 2e - q$. Thus, the probability that $e(n) = e$ is equal to the probability that, for any value of q, we have $q(n) = q$ and $q(n-1) = 2e - q$. If the value of $q(n)$ is independent of the value of $q(n-1)$, as is theoretically expected for sufficiently small values of $\tilde{q}$, then this probability is equal to the probability that $q(n) = q$ times the probability that $q(n-1) = 2e - q$. Furthermore, since the probability densities of $q(n)$ and $q(n-1)$ are each equal to $P_1(q)$, this probability is equal to $P_1(q)P_1(2e - q)$. The probability density distribution of the values of $e_2(n)$, $P_2(e)$, then is the integral over all values of q. That is,

$$P_2(e) = 2 \int_{-\frac{1}{2}\tilde{q}}^{\frac{1}{2}\tilde{q}} P_1(q)P_2(2e - q)dq$$

Note that $P_2(e)$ is essentially equal to the convolution of $P_1(q)$ with itself. For $P_1(q)$ as shown in Figure 13.4, the resulting normalized graph of $P_2(e)$ is the triangle as shown in Figure 13.5. If adjacent samples, $q(n)$ and $q(n-1)$, are not independent, then we would not expect the experimentally obtained probability graph to be a triangle as shown in Figure 13.5.

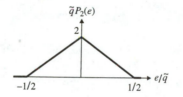

FIGURE 13.5 The probability density of $e_2(n)$

Experimentally obtain the probability bar graphs of $e_2(n)$ using signals s_1, s_2, s_3, s_4, and s_r for $D = 1$ and 2. For which cases are the values of $q(n)$ uniformly distributed but adjacent samples are not independent? Also, for which cases are the values of $q(n)$ uniformly distributed and also adjacent samples are independent?

Now experimentally determine the minimum value of D for which the values of $q(n)$ are uniformly distributed and adjacent samples are independent using any signal except, of course, s_c. From these results, what would you expect the result to be if $s(t)$ were composed of the sum of many sinusoids? Thus, what would you expect the result to be if $s(t)$ were a speech waveform?

■ QUESTION 13.9   A speech waveform for which $|s(t)| \leq 1$ is to be the input of an A/D converter. It is desired that the samples of the quantization error be uniformly distributed

and independent of one another. Use your experimental results to determine the
smallest required word length.

To summarize, this experiment has demonstrated that an analytically difficult problem can sometimes be circumvented by the creation of a statistically
equivalent model. For our case, we avoided the difficult nonlinear analysis of a
quantizer. Also, the statistically equivalent model we obtained is that depicted
by Figure 13.2 in which, under the conditions established in this experiment,
the values of $q(n)$ are uniformly distributed between $-\frac{1}{2}\tilde{q}$ and $\frac{1}{2}\tilde{q}$ and are independent of one another. Such sequences are easily obtained using the random
number generator available on most computers. We knew that such a model was
valid from theory, but the largest quantization interval, $\tilde{q}$, for which the model is
valid had to be determined experimentally. This is an example of how theory and
experiment complement each other. With this model, the effect of A/D quantization error on the system output can be studied statistically without concern for
the specific input signal, $s(t)$.

AN ILLUSTRATION

The MATLAB script that appears when you run `expr13` is shown below.

EXPERIMENT 13

```
FOR THE STUDY OF PROBABILITY DISTRIBUTION OF THE A/D QUANTIZATION ERROR,
SAMPLES OF A SIGNAL (TO BE CHOSEN BELOW) ARE QUANTIZED TO D DECIMALS.
THE ONLY ACCEPTABLE VALUES OF D ARE 0 <= D <= 8.
   Desired value of D =
```

After you enter the value of D, the following script for specifying the parameter
N appears:

```
THE SIGNAL IS SAMPLED AT THE TIME INSTANTS t = 1, 2, ..., N
   Desired value of N =
```

The script then provides information about six input signals.

```
YOU HAVE SIX INPUT SIGNAL CHOICES:
    sc(t) = a constant of your choice between -3.2 and +3.2
    s1(t) = sin(t/11)
    s2(t) = (1/2)*{sin(t/11)+cos(t/79)}
    s3(t) = (1/3)*{sin(t/11)+cos(t/79)+sin(t/31)}
    s4(t) = (1/4)*{sin(t/11)+cos(t/79)+sin(t/31)+cos(t/57)}
    sr(t) = a quasi-random sequence uniformly distributed between -1 and +1
*** PRESS ENTER TO CONTINUE ***
```

After you press the Enter key, the MATLAB menu window shown below appears, from which the desired signal $s(t)$ can be chosen.

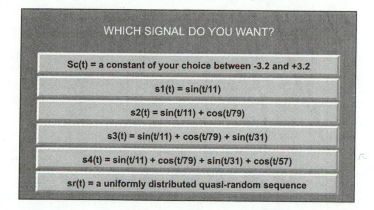

After you complete the sampling quantization error analysis, a MATLAB figure window appears. It contains two graphs: the top graph shows the normalized error distribution of $e_1(n)$, while the bottom graph shows the same for the error $e_2(n)$.

For example, the above two graphs obtained using signal $s_1(t)$ with $D = 0$ and a sample size of 100,000 are shown in Figure 13.6. Note that all the data are given in the top part of each graph. The mean is the average value of the samples of the quantization error and sigma is the standard deviation (which is the rms value) of the samples of the quantization error. For your assistance, the heights of the smallest and largest bars are also given.

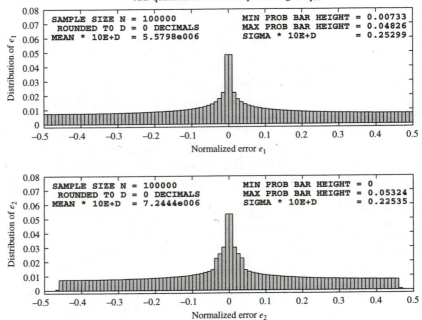

FIGURE 13.6   Plots of probability graphs in Experiment 13

Experiment 14

MULTIPLICATION QUANTIZATION ERROR

BACKGROUND

We shall examine multiplication quantization error in this experiment. The effect of this error on the output of a discrete system will be examined in Experiments 15 and 16.

Standard computers are designed to store and process numbers in two different forms: fixed point and floating point. The floating-point form can represent a much larger dynamic range of numbers than the fixed-point form because it uses a variable accuracy in representing a number. Since the floating-point form is an exponential form corresponding to scientific notation, the quantization error increases with increasing number size. In the fixed-point form, however, the quantization error is independent of number size. In this experiment, we shall study only the quantization error introduced by a computer using fixed-point arithmetic.

A number is represented with a fixed number of decimal places in fixed-point arithmetic. Thus, in a computer using fixed-point arithmetic with three decimal places, the value of π is stored as 3.142 for which the resulting quantization error is

$$3.142 - \pi = 0.00040734641\ldots$$

In such a computer, all numbers are stored with exactly three decimal places. Additional error is incurred if an operation on a stored number results in a number with more than three decimal places that then must be quantized to three decimal places for storage in a register.

In simulating a linear, shift-invariant (LSI) discrete system on a computer, the only operations performed on the stored numbers are delay, addition, and multiplication by a constant. The operation of delay is essentially just the transference of a number from one storage register to another. This operation thus does not result in additional error. The addition of two numbers with D decimal places is again a number with D decimal places, so addition also does not result in additional error. It is possible, of course, for the sum to be so large that overflow occurs. However, overflow usually can be avoided with proper programming and so is not considered in our study.

On the other hand, multiplication can introduce additional error since the product of two numbers with D decimal places is a number with $2D$ decimal places. The result stored by the computer is the product rounded to D decimal places. Thus, in a computer using fixed-point arithmetic with $D = 3$, the square of π is 9.872, which is the square of 3.142 rounded to three decimal places. Be careful not to confuse this with the actual square of π rounded to three decimal places, which is 9.870. The additional error incurred by the squaring operation is

$$9.872 - (3.142)^2 = -0.000164$$

This additional error is called the *multiplication quantization error*. Please note that the multiplication of two different numbers does not result in a multiplication quantization error if one of the numbers is an integer. It is possible, of course, for the product to be so large that overflow occurs. However, as we stated, overflow usually can be avoided with proper programming and so is not considered in our study.

In Experiment 13, a statistically equivalent model was obtained for the A/D quantization error that greatly simplifies the study of the effects of this error in a LSI discrete system. Can such a model also be obtained for multiplication quantization error? If you have not done Experiment 13 on A/D quantization, it would be helpful at this point to read the statistical concepts presented there since they will be useful in understanding this experiment. The system we desire to model is shown in Figure 14.1, in which $x(n)$ is a sequence quantized to D decimal places and c is a constant. One objective of this experiment is to determine conditions for which this system can be modeled as shown in Figure 14.2, in which $q(n)$ is independent of $x(n)$ and is statistically equivalent to the multiplication quantization error, $e(n)$. When such a model is valid, the nonlinear system shown in Figure 14.1 can be replaced with the statistically equivalent linear system of Figure 14.2. Superposition then can be used to study the statistical effects of multiplication quantization error. This study is the subject of the next two experiments.

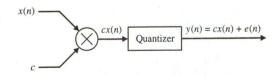

FIGURE 14.1   System for multiplication quantization error

Another objective of this experiment is to determine the required statistical properties of $q(n)$. We would like them to be the same as in the model for A/D conversion. That is, we would like the values of $q(n)$ to be uniformly distributed between $\pm\tilde{q}/2$ (where $\tilde{q}$ is a quantization interval) and also $q(n)$ to be statistically independent of $q(m)$ for $m \neq n$. Through this experiment, it will be seen that this model is valid with, however, some important restrictions.

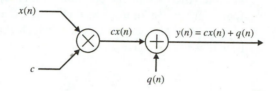

FIGURE 14.2   Linear system model for multiplication quantization error

To better understand multiplication quantization error, consider the computer to use fixed-point arithmetic with D decimal places. Clearly, the statistical model is not valid if c is an integer because then, as discussed above, there is no multiplication quantization error. Now consider the case in which $c = 0.10$. The product $cx(n)$ is obtained by shifting the decimal point of $x(n)$ just one place to the left. If the last integer were just dropped to store the product, the error would then simply be

$$e(n) = k \cdot 10^{-(D+1)}$$

in which k is the integer dropped. However, the product is rounded to D decimal places so that the multiplication quantization error is

$$e(n) = p \cdot 10^{-(D+1)}$$

in which p is an integer between -5 and 5. For example, consider the case in which $D = 3$ and $x(n) = 0.k_1k_2k_3$ in which k_1, k_2, and k_3 are integers. Then, $cx(n) = 0.0k_1k_2k_3$, and if the product were quantized by rounding up, the error,

$$e(n) = \begin{cases} 0.000k_3 & \text{for} \quad k_3 < 5 \\ 0.001 - 0.000k_3 & \text{for} \quad k_3 \geq 5 \end{cases}$$

To illustrate this, if $x(n) = 0.123$, then with $c = 0.100$, the quantized value of $cx(n)$ is 0.012 and the error, $e(n)$, is

$$e(n) = 0.0123 - 0.012 = 0.0003$$

However, if $x(n) = 0.126$, then with $c = 0.100$, the quantized value of $cx(n)$ is 0.013 and the error, $e(n)$, is

$$e(n) = 0.013 - 0.0126 = 0.0004 = 0.001 - 0.0006$$

We thus note that the multiplication quantization error for this case is only *one* of ten discrete values between $\pm\tilde{q}/2$ so it clearly cannot be uniformly distributed.

The complete theoretical analysis of the multiplication quantization error is rather involved and, to our knowledge, has not been done. We are not aware of any theoretical result, as there is in the case of A/D conversion, that can be used to guide us in determining the conditions for which the model of Figure 14.2 is valid. After writing the program for this experiment, we have experimentally found certain conditions, which we shall examine, for which the model appears to be valid. This again is a case in which experiment and theory complement each other.

EXPERIMENTAL PROCEDURE

■

The MATLAB function for this experiment is `expr14`. After running `expr14`, you will be given three input choices: the number of decimal places D, the value of the constant C, and the number of samples N. A MATLAB window will open to provide several choices for the signal $s(t)$. The chosen signal is sampled at times $t = nT_s = 0, 1, 2, 3, \ldots$, and N samples are used for quantization error analysis. The signal choices are

- $s_c(t) =$ a positive constant of your choice between 0 and 1.6
- $s_1(t) = \sin(t/11)$
- $s_2(t) = \frac{1}{2}\{\sin(t/11) + \cos(t/79)\}$
- $s_3(t) = \frac{1}{3}\{\sin(t/11) + \cos(t/79) + \sin(t/31)\}$
- $s_4(t) = \frac{1}{4}\{\sin(t/11) + \cos(t/79) + \sin(t/31) + \sin(t/57)\}$
- $s_r(t) =$ a pseudo-random sequence uniformly distributed between -1 and 1

After you complete the multiplication quantization error analysis, a MAT-LAB figure window will open; the window will contain two graphs. The graph displayed in the top half of the MATLAB figure window is a normalized bar graph of the fraction of samples of $e(n)$ [referred to as $e_1(n)$] that lie in each $0.01\tilde{q}$-wide interval. The graph displayed in the bottom half is a normalized bar graph of the fraction of the sum of samples $e(n)$ and $e(n-1)$ [referred to as $e_2(n)$] that lie in each $0.001\tilde{q}$-wide interval. All the pertinent statistical information is displayed along the top of each graph.

As we discussed, there are certain values of the constant, c, for which there is only a discrete set of values of $e(n)$ so that the multiplication quantization error cannot be uniformly distributed. From our earlier discussion, we clearly expect this to be the case if $c = 0.k$ in which k is an integer. First check this expectation for various values of D by choosing the signal to be $s_r(t)$, a random sequence as described in the screen illustration. Choose the sample size to be 3,000 because the sample does not have to be very large for this determination.

■ QUESTION 14.1   Discuss why the signal $s_r(t)$ was chosen for this case. Is there only a discrete set of values of $e(n)$ for any values of D and k? Verify your experimental results theoretically.

Now examine the case for which $c = 0.k_1k_2$ in which k_1 and k_2 are integers. First choose the case in which $k_1 = k_2 = k$. With the same parameters as above, determine the values of k for which there is only a discrete set of values of $e(n)$. Now experimentally examine the cases for which $k_1 = k$ and $k_2 = k \pm p$ for $p = 1, 2,$ and 3.

■ QUESTION 14.2   For which cases is there only a discrete set of values of $e(n)$? One value is $c = 0.55$ and $D = 5$. Experimentally examine this case and explain why the set of values of $e(n)$ is discrete for this case.

Even though the values of $e(n)$ are not a discrete set, the distribution still may not be uniform for any value of D. For example, let $c = 1/9$ rounded to D decimal places and let the signal be $s_1(t)$, a sinusoidal signal as described in the screen illustration. Examine the distribution of the values of $e(n)$ for various values of D using a sample size of at least 40000. What are the most likely values of the multiplication quantization error for this case? Now repeat with $c = 1/3$.

We therefore note that there are many situations in which the proposed model is not valid. From our own experimental studies, it appears, however, that the model is essentially valid for any signal if

$$c = k.k_1k_2 \ldots k_p$$

in which the k's are nonzero integers and $3 \leq p \leq (D-2)$. By *essentially valid*, we mean that the statistics of the values of $e(n)$ *closely approximate* those of the model. Check this by first determining whether the distribution is uniformly distributed between $\pm \tilde{q}/2$ and then determining whether $e(n)$ is independent of $q(m)$ for $m \neq n$ by determining whether the distribution is a triangle for $L = 2$. (See the discussion in Experiment 13 on A/D quantization for an explanation of why this is a reasonable test for independence.) To allow you the greatest freedom in your study of multiplication quantization error, we have not quantized the constant, c. It is used with the full number of decimal places you enter. Only the chosen sequence, $x(n)$, and the product, $c\,x(n)$, are quantized to D decimal places. Discuss your experimental methodology and the specific cases you studied for an examination of our experimental result. Your discussion should include your reasons for choosing the specific cases used in your examination and the criterion you used to determine whether the statistics are close to those of the model.

AN ILLUSTRATION

The desired signal is chosen from the menu listed on the first screen shown below.

EXPERIMENT 14

FOR THE STUDY OF PROBABILITY DISTRIBUTION OF MULTIPLICATION QUANTIZATION
ERROR, SAMPLES OF A SIGNAL (TO BE CHOSEN BELOW) ARE QUANTIZED TO D DECIMALS
THE ONLY ACCEPTABLE VALUES OF D ARE 0 <= D <= 8.
Desired value of D = 5

THE QUANTIZED SIGNAL IS MULTIPLIED BY A CONSTANT. ONLY VALUES <= 2 ARE
ACCEPTED.
Value of the constant C = 0.11111

THE QUANTIZED PRODUCT IS SAMPLED AT THE INSTANTS t = 1, 2, ..., N
Desired value of N = 10000

YOU HAVE SIX INPUT SIGNAL CHOICES:

    sc(t) = a positive constant of your choice between 0 and +1.6
    s1(t) = sin(t/11)
    s2(t) = (1/2)*{sin(t/11)+cos(t/79)}
    s3(t) = (1/3)*{sin(t/11)+cos(t/79)+sin(t/31)}
    s4(t) = (1/4)*{sin(t/11)+cos(t/79)+sin(t/31)+cos(t/57)}
    sr(t) = a quasi-random sequence uniformly distributed between -1 and +1

*** PRESS ENTER TO CONTINUE ***

After you press the Enter key, the MATLAB menu window shown below appears, from which the desired signal $s(t)$ can be chosen.

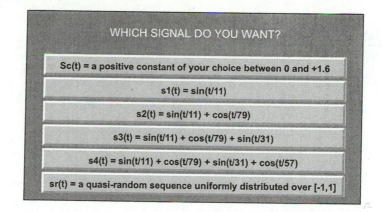

A MATLAB figure window that contains the multiplication quantization error distributions will appear after calculations are completed. An example graph for signal $s_1(t)$ with $D = 5$ and $c = .11111$ is shown in Figure 14.3. Note that all the data are printed at the top of the graph. The mean is the average and sigma is the standard deviation (or rms value) of the values of the multiplication quantization error samples. For convenience, the smallest and largest values of the probability bars are listed.

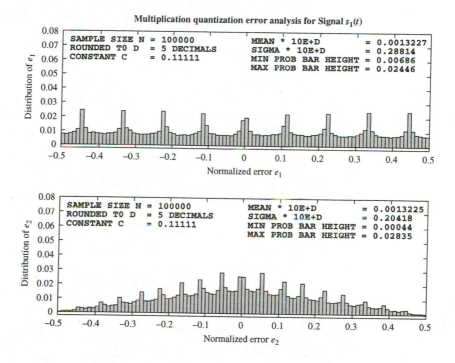

FIGURE 14.3   Plots of probability graphs in Experiment 14

Experiment 15

MULTIPLICATION QUANTIZATION ERROR IN DF1 REALIZATION

BACKGROUND

As discussed in the introduction to this chapter, various sequences are multiplied by constants in the realization of a LSI discrete system. Each multiplication can result in a multiplication quantization error in accordance with the results of Experiment 14. The contribution of this error to the output of a LSI discrete system will be investigated in this experiment. Generally, an exact theoretical analysis of the resulting output error is not possible. We thus will investigate in this experiment whether a reasonable theoretical estimation of the error can be obtained or, if not, whether a simple theoretical upper or lower bound of the error can be obtained. The objective of this experiment is to better understand multiplication quantization error in LSI discrete systems.

We begin by considering a LSI system with the input $x(n)$ and output $y(n)$ described by eq. (15.1):

$$y(n) = b_0 \, x(n) + b_1 \, x(n-1) + a_1 \, y(n-1) \tag{15.1}$$

Each of the three multiplications by the constants b_0, b_1, and a_1 can result in a multiplication quantization error. Denote the multiplication quantization error resulting from the product $b_0 \, x(n)$ by $e_{b_0}(n)$, that from the product $b_1 \, x(n-1)$ by $e_{b_1}(n-1)$, and that from the product $a_1 \, y(n-1)$ by $e_{a_1}(n-1)$. The *Direct Form 1* (DF1) realization of the system described by eq. (15.1) then is as shown in Figure 15.1.

Note that the point at which the errors are added can be moved to the position shown in Figure 15.2 without affecting the output, $y(n)$. The contribution of the three multiplication quantization errors to the output $y(n)$ is thus seen to be the output of an autoregressive filter with a single pole at $z = a_1$ and with an input equal to the sum of the three multiplication quantization errors. Thus, if

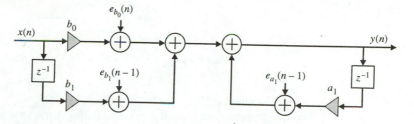

FIGURE 15.1 Direct Form 1 realization of the LSI system in eq. (15.1) with multiplication quantization errors introduced

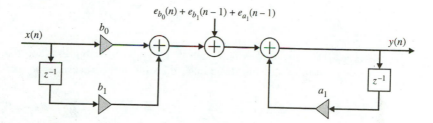

FIGURE 15.2 An equivalent system to the Direct Form 1 realization in Figure 15.1

we let $e(n)$ be the contribution of the error to the output $y(n)$, then Figure 15.3 is a model of the error, $e(n)$.

Two quantities of importance in characterizing the error, $e(n)$, are its mean, $\bar{e}$, and its standard deviation, σ_e. The mean of e, $\bar{e}$, is simply its average value. The standard deviation, σ_e, is the square root of the variance, σ_e^2, which is defined as

$$\sigma_e^2 = \overline{(e - \bar{e})^2} \qquad (15.2)$$

That is, the variance is equal to the average of the square of the deviation of the values of e from its mean. Thus, the standard deviation, σ_e, is a measure of the average deviation of e from its mean.

These quantities of $e(n)$ could be determined theoretically if certain statistical averages of the filter input were known. To determine the required specific information about the input sequence, we shall theoretically determine the mean,

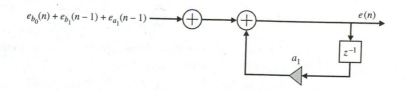

FIGURE 15.3 An equivalent system used to study multiplication quantization error

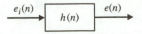

FIGURE 15.4   A LSI system with impulse response $h(n)$

$\bar{e}$, and variance, σ_e^2, of $e(n)$ for the general case of a LSI system with the unit sample response $h(n)$, input $e_i(n)$, and output $e(n)$, as shown in Figure 15.4.

We begin by expressing the system output as the convolutional sum of the system unit sample response and the system input

$$e(n) = \sum_{k=-\infty}^{\infty} h(k)e_i(n-k) \tag{15.3}$$

The mean of $e(n)$, $\bar{e}$, can be obtained by using the fact that the average of the sum of variables is equal to the sum of their averages so that

$$\bar{e} = \overline{e(n)} = \sum_{k=-\infty}^{\infty} h(k)\overline{e_i(n-k)} = \overline{e_i} \sum_{k=-\infty}^{\infty} h(k) = \overline{e_i} \ H(z)|_{z=1}$$

$$= \overline{e_i} \ H(1) \tag{15.4}$$

The result in eq. (15.4) states the expected result that the d-c output is equal to the d-c input times the system d-c gain. For the cases of interest to us in this calculation, the d-c input, $\overline{e_i}$, is zero so that, from eq. (15.4), $\bar{e} = 0$. Consequently, from eq. (15.2), we obtain

$$\sigma_e^2 = \overline{e^2} \tag{15.5}$$

Now, the square of $e(n)$ can be expressed in terms of the input, $e_i(n)$, in the following manner:

$$e^2(n) = \left[\sum_{k=-\infty}^{\infty} h(k)e_i(n-k) \right]^2$$

$$= \sum_{k_1=-\infty}^{\infty} h(k_1)e_i(n-k_1) \cdot \sum_{k_2=-\infty}^{\infty} h(k_2)e_i(n-k_2)$$

$$= \sum_{k_1=-\infty}^{\infty} \sum_{k_2=-\infty}^{\infty} h(k_1)h(k_2)e_i(n-k_1)e_i(n-k_2) \tag{15.6}$$

Using the fact that the average of the sum of variables is equal the sum of their averages, we then have from eq. (15.6)

$$\overline{e^2} = \overline{e^2(n)} = \sum_{k_1=-\infty}^{\infty} \sum_{k_2=-\infty}^{\infty} h(k_1)h(k_2)\overline{e_i(n-k_1)e_i(n-k_2)} \tag{15.7}$$

The average under the double summation is called the *autocorrelation* function of $e_i(n)$. A case of interest to us is that for which $e_i(n)$ is independent of $e_i(m)$ for $n \neq m$. For such a case,

$$\overline{e_i(n-k_1)e_i(n-k_2)} = \overline{e_i^2}\,\delta(k_1 - k_2) = \begin{cases} 0 & \text{for } k_1 \neq k_2 \\ \overline{e_i^2} & \text{for } k_1 = k_2 \end{cases} \tag{15.8}$$

With the use of eqs. (15.5) and (15.8), we obtain for eq. (15.7)

$$\sigma_e^2 = \overline{e^2} = \overline{e_i^2} \sum_{k=-\infty}^{\infty} h^2(k) \tag{15.9}$$

We will illustrate these theoretical results with the system described by eq. (15.1). For this, we consider the case for which the values of the multiplication quantization error sequence, $q(n)$, are uniformly distributed between $\pm \tilde{q}/2$ and $q(m)$ is independent of $q(n)$ for $m \neq n$. In Experiment 14, this was found to be a good model for the multiplication quantization error if the number of nonzero decimal places of the multiplying constant is ≥ 3 but $\leq (D-2)$ for a computer using fixed-point arithmetic with D decimal places. For such cases, the multiplication quantization error is uniformly distributed between $\pm \tilde{q}/2$, in which $\tilde{q} = 10^{-D}$, so the probability density distribution of the error is

$$P_q(q) = \begin{cases} 1/\tilde{q} & \text{for } -\tilde{q}/2 \leq q \leq \tilde{q}/2 \\ 0 & \text{otherwise} \end{cases} \tag{15.10}$$

Consequently,

$$\bar{q} = \int_{-\infty}^{\infty} q\,P_q(q)\,dq = \int_{-\tilde{q}/2}^{\tilde{q}/2} \frac{1}{\tilde{q}} q\,dq = 0 \tag{15.11}$$

and

$$\overline{q^2} = \int_{-\infty}^{\infty} q^2 P_q(q)\,dq = \int_{-\tilde{q}/2}^{\tilde{q}/2} \frac{1}{\tilde{q}} q^2\,dq = \frac{1}{12}\tilde{q}^2 \tag{15.12}$$

Now, we showed the system depicted in Figure 15.3 to be a model for the system output error. For that system, it is easy to show that

$$h(n) = a_1^n\,u(n) \tag{15.13}$$

in which $u(n)$ is the unit step function.

■ QUESTION 15.1   Show that the system unit sample response is as given by eq. (15.13).

Thus, from eq. (15.9),

$$\sigma_e^2 = \overline{e_i^2} \sum_{k=0}^{\infty} a_1^{2n}$$

$$= \overline{e_i^2}\, \frac{1}{1 - a_1^2}, \quad \text{for } |a_1| < 1 \tag{15.14}$$

We now note that

$$e_i(n) = e_{b_0}(n) + e_{b_1}(n-1) + e_{a_1}(n-1) \tag{15.15}$$

so that

$$\overline{e_i} = \overline{e_i(n)} = \overline{e_{b_0}(n)} + \overline{e_{b_1}(n-1)} + \overline{e_{a_1}(n-1)} = 0 \tag{15.16}$$

The average is zero since, from eq. (15.11), each of the averages on the right side of the equation is zero. Now,

$$\overline{e_i^2} = \overline{e_i^2(n)} = \overline{\left[e_{b_0}(n) + e_{b_1}(n-1) + e_{a_1}(n-1) \right]^2}$$

$$= \overline{e_{b_0}^2(n)} + \overline{e_{b_1}^2(n-1)} + \overline{e_{a_1}^2(n-1)} \tag{15.17}$$

$$= \left[\frac{1}{12} + \frac{1}{12} + \frac{1}{12} \right] \tilde{q}^2 = (\tfrac{1}{4}) 10^{-2D} \tag{15.18}$$

To obtain eq. (15.17), it was assumed that the various multiplication quantization errors are independent so that, with eq. (15.11), the average of their products is zero. Equation (15.18) was then obtained using eq. (15.12). *The assumption that the various multiplication errors are independent is not necessarily true.* But without it, the calculation could not be completed since the statistics of the products are not known. One objective of this experiment is to determine whether the error introduced by this assumption is not too large so that it can provide a reasonable bound of the mean-square error. Combining eqs. (15.14) and (15.18), we finally obtain

$$\sigma_e^2 = \frac{10^{-2D}}{4(1 - a_1^2)} \tag{15.19}$$

so, with the independence assumption, the standard deviation of the error, $e(n)$, is

$$\sigma_e = \frac{10^{-D}}{2\sqrt{1 - a_1^2}} \tag{15.20}$$

■ QUESTION 15.2   Equation (15.20) is valid if the various multiplication quantization errors are independent as stated following eq. (15.18). Determine a similar expression for the standard deviation for the other extreme case in which the various multiplication quantization errors are unity correlated so that

$$\overline{e_{b_0}(n)\, e_{b_1}(n-1)} = \overline{e_{b_0}(n)\, e_{a_1}(n-1)} = \overline{e_{b_1}(n-1)\, e_{a_1}(n-1)} = \frac{1}{2}\tilde{q}^2 \tag{15.21}$$

EXPERIMENTAL PROCEDURE

---■---

The MATLAB function for this experiment is `expr15`. After running `expr15`, you will be given two input choices: the number of decimal places D and the number of samples N. A MATLAB window will then open to provide two choices for the signal $s(t)$. The chosen signal is sampled at times $t = nT_s = 0$, 1, 2, 3, ..., and N samples are used for quantization error analysis. The signal choices are

- $s_1(t) = \sin(t/79)$
- $s_r(t)$ = a pseudo-random sequence uniformly distributed between -1 and 1

Finally, a MATLAB command window will prompt you to enter the coefficients b_0, b_1, and a_1 of the DF1 realization.

After you complete the multiplication quantization error analysis, a MATLAB figure window will display a graph. This graph is a histogram of the error, $e(n)$. The error range is divided into 100 intervals. The height of a bar is equal to the fraction of the number of samples of $e(n)$ that lie in that range. The bars are numbered from -50 to 50. The bar width and other statistical information are displayed at the top of each graph.

Note that the DF1 realization can be viewed as a moving average (MA) filter followed in tandem by an autoregressive (AR) filter. We begin by investigating each filter separately. The program will experimentally determine the mean, standard deviation, and distribution of the output multiplication error for each set of coefficients. As in Experiment 14, the coefficient constants are not quantized in the program before multiplication. Thus, be sure you enter them with the exact desired number of decimal places. It is not necessary to enter final zeros. For example, it is not necessary to enter 0.125000 for 1/8; it can be entered as 0.125. For good results, choose the number of samples used by the computer to be at least 100,000.

PART 1

We examine the output multiplication error of just the MA filter in this part. For this, choose $a_1 = 0$ so that

$$y(n) = b_0 x(n) + b_1 x(n-1)$$

Use $D = 6$. Run the program for each of the following three cases:

1. $b_0 = 1$; $b_1 = 1$
2. $b_0 = 0.1111$; $b_1 = 1$
3. $b_0 = 0.1111$; $b_1 = 0.1111$

Explain the results obtained. For this, it would be helpful to review Experiment 14.

PART 2

We examine the output multiplication error of just a single pole AR filter in this part. For this, choose $b_0 = 1$ and $b_1 = 0$ so that

$$y(n) = x(n) + a_1 y(n-1)$$

Use $D = 6$. Run the program for each of the following cases:

$$a_1 = 0.0111, \ 0.3111, \ 0.5, \ 0.5111, \ 0.9911, \ 0.9999$$

Compare the standard deviation for each case with that given by eq. (15.14). Note that $e_{b_0}(n)$ and $e_{b_1}(n)$ are zero in this part. Discuss the changes in the shape and values of the error distribution for increasing values of a_1.

PART 3

To note the effect of the quantized number of decimal places of the coefficients, rerun Parts 1 and 2 with $D = 6$ and the coefficients truncated to one decimal place. Discuss any differences from your previous results.

PART 4

To note the effect of the quantized number of decimal places, rerun Parts 1, 2, and 3 with the coefficients as given but with $D = 2$. Discuss any differences from your previous results.

PART 5

Up to this point, we have considered only a single-pole AR filter. We now examine the effect of adding a pole so that eq. (15.1) is given by

$$y(n) = b_0 x(n) + b_1 x(n-1) + a_1 y(n-1) + a_2 y(n-2) \qquad (15.22)$$

The MATLAB function `expr15` is designed to allow you to input the additional coefficient a_2. Repeat Parts 2, 3, and 4 with $a_1 = 2\alpha$ and $a_2 = -2\alpha^2$ for

$$\alpha = 0.01, \ 0.11, \ 0.22, \ 0.45, \ 0.70$$

Note for this part that eq. (15.14) must be modified since $h(n)$ is no longer given by eq. (15.13). Determine $h(n)$ for the case considered in this part and use eq. (15.9) to determine the standard deviation using the independent assumption and for the other extreme case in which the various multiplication quantization errors are unity correlated. Note for this part that

$$e_i(n) = e_{a_1}(n-1) + e_{a_2}(n-2) \qquad (15.23)$$

instead of that given by eq. (15.15).

PART 6

We now consider the complete DF1 filter by choosing the coefficients of the MA filter to be $b_0 = 0.1111$ and $b_1 = 0.1111$. Now rerun Parts 2, 3, 4, and 5 with the given coefficients of the AR filter.

PART 7   All the above results were obtained with the input being a random sequence. Do the results differ for a sinusoidal input? Run the preceding parts using the sinusoidal input and discuss any changes in the statistics and the distribution of the output multiplication quantization error.

■ **QUESTION 15.3**   Have you reached any general conclusions concerning approximations of the maximum error, minimum error, error standard deviation, and shape of the error distribution curve of a DF1 filter? If so, discuss them and your reasoning for them.   ╲

AN ILLUSTRATION
───────────────── ■ ─────────────────

The MATLAB script that appears when you run `expr15` is shown below.

EXPERIMENT 15

```
   FOR THE DISTRIBUTION OF THE DIRECT FORM 1 FILTER OUTPUT ERROR DUE TO
MULTIPLICATION QUANTIZATION, SAMPLES OF A SIGNAL (TO BE CHOSEN BELOW) ARE
QUANTIZED TO D DECIMALS. THE ONLY ACCEPTABLE VALUES OF D ARE 0 <= D <= 8.
 Desired value of D =
```

After you enter the value of D, the following script for specifying the parameter N appears.

```
   THE QUANTIZED PRODUCT IS SAMPLED AT THE INSTANTS t = 1, 2, ..., N
 Desired value of N =
```

The script then provides information about two input signals.

```
YOU HAVE TWO INPUT SIGNAL CHOICES:

   s1(t) = sin(t/79)
   sr(t) = a quasi-random sequence uniformly distributed between -1 and +1

*** PRESS ENTER TO CONTINUE ***
```

After you press the **Enter** key, the following MATLAB menu window appears from which the desired signal $s(t)$ can be chosen.

WHICH SIGNAL DO YOU WANT?

s1(t) = sin(t/79)

sr(t) = a quasi-random sequence uniformly distributed over [-1,1]

After you select the desired input, the MATLAB command window is activated for entering filter coefficients in eq. (15.22). These coefficients are entered in the program as array inputs to MATLAB. The entered quantities for arrays b=[b0,b1] and a=[a1,a2] must be enclosed in square brackets. This is shown below for $b_0 = 0.11111$, $b_1 = 0$, $a_1 = 0.875$, and $a_2 = 0$.

```
YOU HAVE CHOSEN THE SIGNAL s1(t)

THE EQUATION OF A DIRECT FORM 1 REALIZATION IS

  y(n) = b0*x(n) + b1*x(n-1) + a1*y(n-1) + a2*y(n-2)

  SPECIFY THE DESIRED VALUES OF THE FILTER COEFFICIENTS AS VECTORS, e.g.,
        [b0, b1] as an array b
  and   [a1, a2] as an array a
b = [0.11111]
a = [0.875]
```

Note that if the trailing coefficients b_1 or a_2 are zero, then their values need not be entered. The program will assign the zero values.

After you complete the sampling quantization error analysis, a MATLAB figure window appears. It contains the error distribution of $e(n)$. For example, the graph obtained using signal $s_1(t)$ with $D = 5$, a sample size of 100,000, and filter coefficients b=[0.11111,0] and a=[0.875] is shown in Figure 15.5. Note that all the data are given at the top part of the graph. The mean of the error distribution is 1.1483×10^{-7} and the standard deviation is 0.925×10^{-5}. For your assistance, the bar width, the total error range, and the heights of the smallest and largest bars are also given. For example, the total error range is $\pm 3.7602 \times 10^{-5}$ and there are 100 error intervals; hence, the probability bar width is

$$\frac{2(3.7602)10^{-5}}{100} = 0.075204 \times 10^{-5}$$

which is also displayed.

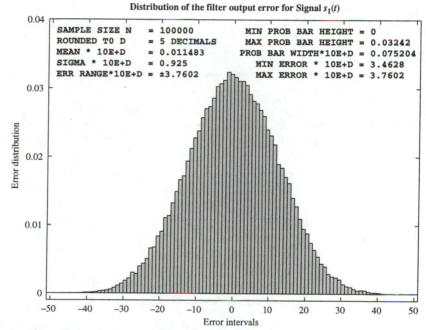

FIGURE 15.5   Plot of probability graph in Experiment 15

Experiment 16

EFFECT OF FILTER REALIZATION ON THE OUTPUT MULTIPLICATION QUANTIZATION ERROR

BACKGROUND

The error at the output of the Direct Form 1 realization of an LSI system was examined in Experiment 15. However, a system can be realized in a number of different forms. Because of their different structures, the contribution of the multiplication quantization error to the output of each form generally is different. An objective of this experiment is to examine these differences and to determine whether a reasonable theoretical estimation of the error can be obtained or whether a simple theoretical upper or lower bound of the error can be obtained. If this is not possible, then at the least we would like to know whether any guidelines can be deduced to determine the realization with the smallest multiplication quantization error.

For this study, the contribution to the output of five different forms of realizing a system with, at most, two poles and two zeros will be investigated. These forms are briefly described below.

1. As discussed in Experiment 15, the equation of the Direct Form 1 (DF1) is

$$y(n) = b_0 x(n) + b_1 x(n-1) + b_2 x(n-2) + a_1 y(n-1) + a_2 y(n-2) \quad (16.1)$$

2. The equations of the second form, called the *Direct Form 2* (DF2), are

$$y(n) = b_0 w(n) + b_1 w(n-1) + b_2 w(n-2) \quad (16.2a)$$

$$w(n) = x(n) + a_1 w(n-1) + a_2 w(n-2) \quad (16.2b)$$

3. The equations of the third form, called the *Parallel Form* (PF), are

$$y(n) = w_0(n) + w_1(n) + w_2(n) \qquad (16.3a)$$

$$w_0(n) = r_0 x(n) \qquad (16.3b)$$

$$w_1(n) = r_1 x(n) + p_1 w_1(n-1) \qquad (16.3c)$$

$$w_2(n) = r_2 x(n) + p_2 w_2(n-1) \qquad (16.3d)$$

in which p_1 and p_2 are poles of the system function $H(z)$, which are assumed to be real-valued; r_1 and r_2 are the residues at the corresponding pole locations, which are then also real-valued; and r_0 is a real-valued constant.

4. The equations of the fourth form, which we refer to as the *Cascade Form 1* (CF1) because it is the cascade (or series) connection of two Direct Form 1 realizations, are

$$y(n) = w_2(n) + p_2 y(n-1) \qquad (16.4a)$$

$$w_2(n) = w_1(n) + c_2 w_1(n-1) + p_1 w_2(n-1) \qquad (16.4b)$$

$$w_1(n) = c_0 x(n) + c_1 x(n-1) \qquad (16.4c)$$

in which p_1 and p_2 are poles of the system function $H(z)$, which are assumed to be real-valued, and c_0, c_1, and c_2 are real-valued constants.

5. Finally, the equations of the fifth form, which we call the *Cascade Form 2* (CF2) because it is the cascade connection of two Direct Form 2 realizations, are

$$y(n) = w_2(n) + c_2 w_2(n-1) \qquad (16.5a)$$

$$w_2(n) = c_0 w_1(n) + c_1 w_1(n-1) + p_2 w_2(n-1) \qquad (16.5b)$$

$$w_1(n) = x(n) + p_1 w_1(n-1) \qquad (16.5c)$$

in which the coefficients are as described above.

■ QUESTION 16.1   Draw a realization for each of the five forms from its equations.

■ QUESTION 16.2   Represent each multiplication quantization error as in Figure 15.1. Now, as we did in obtaining the equivalent system of Figure 15.2, note that the multiplication quantization errors in each moving average section can be moved to the output of that section, and those in each autoregressive section can be moved to the input of that section without affecting the output, $y(n)$. Redraw each of your realizations with each of the multiplication quantization errors moved as we have just described.

EXPERIMENTAL PROCEDURE

■

The MATLAB function for this experiment is `expr16`. After running `expr16`, you will be given two input choices: the number of decimal places D and the number of samples N. A MATLAB window will then open to provide two choices for the signal $s(t)$. The chosen signal is sampled at times $t = nT_s = 0$, 1, 2, 3, . . . , and N samples are used for quantization error analysis. The signal choices are

- $s_1(t) = \sin(t/79)$
- $s_r(t) = $ a pseudo-random sequence uniformly distributed between -1 and 1

Finally, a MATLAB command window will prompt you to enter the coefficients of all five realizations. These coefficients should be entered as row vectors, which are described in the next section.

After you complete the multiplication quantization error analysis, five MATLAB figure windows will display the respective bar graphs for the five realizations. These graphs are histograms of the error, $e(n)$. The error range is divided into 100 intervals. The height of a bar is equal to the fraction of the number of samples of $e(n)$ that lie in that range. The bars are numbered from -50 to 50. The bar width and other statistical information are displayed at the top of each graph.

PART 1

In this part, we shall examine all five realizations of four different all-pass discrete filters theoretically. An all-pass filter is one for which the system gain does not vary with frequency. It is generally used to modify the phase-shift of a given system. (See Experiment 10 for a more detailed discussion of all-pass filters.) The system function of such a filter with two poles and two zeros is

$$H(z) = \frac{(z - 1/p_1)(z - 1/p_2)}{(z - p_1)(z - p_2)} \tag{16.6}$$

We shall study four different cases:

1. $p_1 = 0.25$ and $p_2 = -0.25$
2. $p_1 = 0.95$ and $p_2 = -0.95$
3. $p_1 = 0.25$ and $p_2 = 0.1875$
4. $p_1 = 0.90$ and $p_2 = 0.675$

First, derive the constant coefficients of the difference equations required for each of the five realizations and for each of the four cases. You can simplify your computations by noting that $p_2 = -p_1$ in the first two cases and that

$p_2 = 0.75\, p_1$ in the third and fourth cases. Then derive the mean, $\bar{e}$, and the standard deviation, σ_e, for each of the five realizations and for each of the four cases using the same assumptions we used in the derivation of σ_e given by eq. (15.20) in Experiment 15 and in the question following it.

PART 2

Now run the program for this experiment for each of the cases with $D = 8$ and the input $s_r(t)$. Enter each coefficient with six decimal places. The program will experimentally determine the mean and the standard deviation of the output error, $e(n)$, and its distribution for five realizations sequentially. As in Experiments 14 and 15, the coefficient constants are not quantized in the program before multiplication. Thus, make certain you enter them with the exact desired number of decimal places. It is not necessary to enter final zeros. For example, it is not necessary to enter 0.125000 for 1/8; it can be entered as 0.125. For good results, choose the number of samples used by the computer to be at least 100,000.

Discuss your calculated and experimental results. Specifically:

1. If you have done Part 1, compare your calculated values with those obtained experimentally. Are your calculations consistently pessimistic or optimistic?

2. For each case, order the realizations relative to the experimentally obtained standard deviation of the error. Is the standard deviation of the error smallest (largest) for one type of realization for all cases? If you have done Part 1, do your calculations at least predict the filter realization with the smallest (largest) output error standard deviation?

3. In Experiment 15, insight was obtained concerning the output multiplication quantization error of a MA and an AR filter. With that insight, what would you expect the order of increasing standard deviation of the realization to be? Discuss your reasoning for your choice of order and how it compares with the actual order obtained.

4. Discuss the graphs of the output error distributions. Are certain errors much more likely than adjacent values of the error? What does this imply about the assumptions you made in Parts 2 and 3?

PART 3

How do your results change for other values of D? Run at least one of the cases above with $D = 6$ and with the coefficients entered with four decimal places. Discuss your results and compare them with those obtained earlier.

PART 4

To note the effect of the number of decimal places of the coefficients, rerun the case you ran with $D = 6$ but with the coefficients entered with one and two decimal places. Discuss the effect of the number of decimal places of the coefficients on the statistics and the distribution of the output error.

PART 5   Many other studies are possible with the program. For example, all the earlier
                results were obtained with the input being a random sequence. Do they differ for
                other inputs? You can experiment using the other input that is available in the
                program. Also, it is interesting to observe the error statistics for smaller values
                of D.

PART 6   From your studies in this experiment, discuss your conclusions concerning
                the realization to use for the smallest output multiplication quantization error.
                Specifically, does one realization consistently result in the smallest error? If
                not, can you suggest any guidelines for choosing the realization that results in
                the smallest error? That is, given the desired poles and zeros of a LSI discrete
                system, can you suggest any guidelines for choosing the realization that results
                in the smallest output multiplication quantization error?

AN ILLUSTRATION

                For the purpose of this illustration, we will consider a linear system with poles
                at $p_1 = 0.25$ and $p_2 = -0.25$. For this system, the coefficients of the five
                realizations are

DF1 and DF2: $b_0 = 1$, $b_1 = 0$, $b_2 = -16$, $a_1 = 0$, $a_2 = 0.0625$
        PF: $r_0 = 256$, $r_1 = -127.5$, $r_2 = -127.5$, $p = 0.25$, $p_2 = -0.25$
CF1 and CF2: $c_0 = 1$, $c_1 = -4$, $c_2 = 4$, $p_1 = 0$, $p_2 = 0.0625$

                The MATLAB script that appears when you run `expr16` is shown below.

                EXPERIMENT 16

  FOR THE DISTRIBUTION OF FILTER OUTPUT ERROR DUE TO MULTIPLICATION
QUANTIZATION, SAMPLES OF A SIGNAL (TO BE CHOSEN BELOW) ARE QUANTIZED
TO D DECIMALS. THE ONLY ACCEPTABLE VALUES OF D ARE 0 <= D <= 8.
  Desired value of D =

                After you enter the value of D, the following script for specifying the parameter
                N appears.
                The script then provides information about two input signals.

YOU HAVE TWO INPUT SIGNAL CHOICES:

   s1(t) = sin(t/79)
   sr(t) = a quasi-random sequence uniformly distributed between -1 and +1

*** PRESS ENTER TO CONTINUE ***

After you press the Enter key, the MATLAB menu window shown below appears from which the desired signal $s(t)$ can be chosen.

WHICH SIGNAL DO YOU WANT?

s1(t) = sin(t/79)

sr(t) = a quasi-random sequence uniformly distributed over [-1,1]

After you selecting the desired input, the MATLAB command window is activated for entering filter coefficients given in eqs. (16.1) through (16.5). These coefficients are entered into the program as array inputs to MATLAB. This is shown below for the linear system with poles $p_1 = 0.5$ and $p_1 = -0.25$.

```
THE EQUATION OF A DIRECT FORM 1 REALIZATION IS
   y(n) = b0*x(n) + b1*x(n-1) + b2*x(n-2) + a1*y(n-1) + a2*y(n-2)

 THE EQUATIONS OF A DIRECT FORM 2 REALIZATION ARE
    y(n) = b0*w(n) + b1*w(n-1) + b2*w(n-2)
    w(n) =     x(n) + a1*w(n-1) + a2*w1(n-2)

  SPECIFY THE DESIRED VALUES OF THE FILTER COEFFICIENTS AS ARRAYS, e.g.,
        [b0, b1, b2] as an array b
     and  [a1, a2] as an array a
b = [1,0,-16]
a = [0,0.0625]

  THE EQUATIONS OF A PARALLEL FORM REALIZATION ARE
     y(n) = w0(n) + w1(n) + w2(n)
    w0(n) = r0*x(n)
    w1(n) = r1*x(n) + p1*w1(n-1)
    w2(n) = r2*x(n) + p1*w2(n-1)

  SPECIFY THE DESIRED VALUES OF THE FILTER COEFFICIENTS AS VECTORS, e.g.,
        [r0, r1, r2] as a vector r
     and  [p1, p1] as a vector p
r = [256, -127.5,-127.5]
p = [0.25,-0.25]

  THE EQUATIONS OF A CASCADE FORM 1 REALIZATION ARE
    w1(n) = c0*x(n) + c1*x(n-1)
    w2(n) = w1(n) + c2*w1(n-1) + p1*w2(n-1)
     y(n) = w2(n) + p2*y(n-1)
```

```
THE EQUATIONS OF A CASCADE FORM 2 REALIZATION ARE
   w1(n) = x(n) + p1*x(n-1)
   w2(n) = c0*w1(n) + c1*w1(n-1) + p2*w2(n-1)
    y(n) = w2(n) + c2*w2(n-1)


SPECIFY THE DESIRED VALUES OF THE FILTER COEFFICIENTS AS VECTORS, e.g.,
        [c0, c1, c2] as a vector c
   and  [p1, p2] as a vector p
c = [1,-4,4]
p = [0.25,-0.25]
```

After you complete the sampling quantization error analysis, five MATLAB figure windows appear, one each for five realizations. They contain the normalized error distributions of $e(n)$. For example, the graph obtained using signal $s_1(t)$ with $D = 5$, a sample size of 100,000, and the Direct Form 2 filter realization is shown in Figure 16.1. Note that all the data are given at the top part of the graph. The mean of the error distribution is 8.8075×10^{-8} and the standard deviation is 4.651×10^{-5}. For your assistance, the bar width, the total error range, and the heights of the smallest and largest bars are also given. For example, the total error range is $\pm 8.995 \times 10^{-5}$ and there are 100 error intervals; hence, the probability bar width is

$$\frac{2(8.995)10^{-5}}{100} = 0.1799 \times 10^{-5}$$

which compares very favorably with the value listed at the top of the graph.

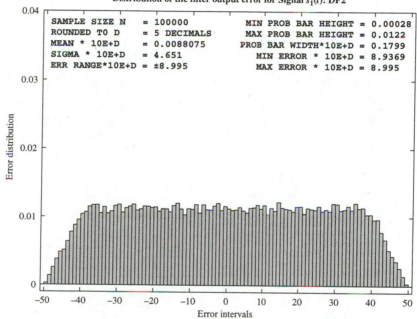

FIGURE 16.1 Plot of probability graph for the parallel form realization in Experiment 16